SCIENCE, TECHNOLOGY AND INNOVATION:
For Sustainable Future in the Global South

SCIENCE, TECHNOLOGY AND INNOVATION:
For Sustainable Future in the Global South

Edited by Mammo Muchie, Amare Desta and Mentesnot Mengesha

AFRICA WORLD PRESS
TRENTON | LONDON | CAPE TOWN | NAIROBI | ADDIS ABABA | ASMARA | IBADAN | NEW DELHI

AFRICA WORLD PRESS
541 West Ingham Avenue | Suite B
Trenton, New Jersey 08638

Copyright © 2016
All rights reserved. No part of this publication may be reproduced, stored in a retrieval system or transmitted in any form or by any means electronic, mechanical, photocopying, recording or otherwise without the prior written permission of the publisher.
Book design: Dawid Kahts
Cover design: Ashraful Haque

Library of Congress Cataloging-in-Publication Data

Names: Muchie, Mammo, editor. | Desta, Amare, 1964- editor, author. |Mengesha, Mentesnot, 1959- editor, author.
Title: Science, technology and innovation : for sustainable future in the global south / edited by Mammo Muchie, Amare Desta and Mentesnot Mengesha.
Other titles: Science, technology and innovation (Africa World Press)
Description: Trenton : Africa World Press, 2016. | Includes bibliographical references and index.
Identifiers: LCCN 2016020314| ISBN 9781569024386 (hb : alk. paper) | ISBN 9781569024393 (pb : alk. paper)
Subjects: LCSH: Technological innovations--Economic aspects--Developing countries. | Technological innovations--Social aspects--Developing countries. | Sustainable development--Developing countries. | Information technology--Developing countries. | Telecommunication--Developing countries. | Technological innovations--Government policy--Developing countries.
Classification: LCC HC59.72.T4 S38 2016 | DDC 338.064091724--dc23 LC record available at https://lccn.loc.gov/2016020314

CONTENTS

LIST OF ILLUSTRATIONS ix
LIST OF FIGURES x
ACKNOWLEDGMENTS xiii
ACRONYMS xv
FOREWORD xix
INTRODUCTION xxi

PART I: INNOVATIVE AND SUSTAINABLE DEVELOPMENTS AS SOCIO-ECONOMIC SOLUTIONS

Chapter 1: Grassroots Innovations for Zimbabwe's Smallholder Agricultural Transformation: Evidence from Informal Metal Industry Survey.. 3
Kingstone Mujeyi

Chapter 2: Indigenous Agricultural Technology in Nigeria: Case study of National Centre for Agricultural Mechanisation................................ 17
O. I. Ogunyemi and A. S. Adedokun

Chapter 3: Rural Innovations and Knowledge Systems Development and Dissemination among cassava cooperative farmers in Southern Nigeria 27
Edwards Adeseye Alademerin

v

Chapter 4: The role of Innovation in Development – *Could Lessons be Drawn for Ethiopia?* 47
Bedru B. Balana

Chapter 5: A Conceptual Framework for Building A Home-grown Public-Private Partnership Platform to deliver public Information Services in Developing Economies 59
Temesgen A. Weseniand and Richard T. Watson

Chapter 6: FDI-Growth Nexus in Ethiopia: Is there any causality?..........75
Yesuf M. Awel and Tsehaye Weldegiorgis

Chapter 7: Scenario planning as a management tool for Sustainable Aquaculture 89
Worku Jimma, Daniel Adjei-Boateng and Nicolas Van Vosselen

Chapter 8: A Study of a Public-Private Acquisition System applying Transaction Cost Economics Principles 99
Ermias Kebede

PART 2: APPROPRIATNESS AND APPLICABILITY OF ICT FOR GROWTH AND DEVELOPMENT

Chapter 9: Assessment of Knowledge Management Practices in Jimma University: Consideration of Technology, Leadership, Organization and Learning Pillars 131
Haftamu Ebuy, Rahel Bekele and Worku Jimma

Chapter 10: Knowledge, Attitude and Utilization of Information Communication Technologies (ICTs) in an Ethiopian Medical Teaching Hospital..........143
Melisachew Adane, Lemma Lessa and Solomon Shiferaw

Chapter 11: Multi-criteria decision modelling for infrastructure development: A case of Ethiopian highway rehabilitation projects 161
Tamirat Fikre Nebiyu

Chapter 12: The Practice of opting for Open Source Solutions in Higher Education Institutions of Ethiopia 177
Florida Alemayehu and Lemma Lessa

Chapter 13: Assessing IT Governance Practices in Ethiopian Financial Institutions 193
Mengistu Bogale

Chapter 14: Hybrid Software Architecture Design Pattern Model 207
Asebe Jeware and Nassir Dino

PART 3: POLICY TOOLS TO UNPACK WHAT DEVELOPING COUNTRIES CURRENTLY FACE

Chapter 15: Exploring the extent of Information and Communication Technology (ICT) in supporting pedagogical practices in developing countries 219
Amare Desta

Chapter 16: Localization in Iranian Context: A Multi-Case Study 241
Mohammad-Ali Shafia, Mehdi Mohammadi, Ali-Reza Babakhan, Mohammad Ameri and Faramarz Lotfali

Chapter 17: Knowledge Management Practices in Development and Humanitarian Aid Organizations in Ethiopia 257
Hermella Ayalew

Chapter 18: A Learning Oriented Approach for Organizational Requirements Development 273
Mesfin Kifle

Chapter 19: Understanding Community Resource-Use Linkages: A Case Study Conducted in Southeast and Southcentral Alaska 285
Mekbeb E. Tessema, Robert J. Lilieholm, Dale J. Blahna, Linda E. Kruger and Joanna Endter-Wada

Chapter 20: Conceptualising Power and Empowerment in the Context of Public Participation in Urban Development Projects in Sub-Saharan African Cities. 313
Mentesnot Mengesha and Mammo Muchie

CONCLUDING REMARKS 327

NOTE ON CONTRIBUTORS 331

INDEX 339

LIST OF ILLUSTRATIONS

List of Tables

Chapter 1
Table 1: Educational Level Attainment 9
Table 2: Source of skills training 11

Chapter 3
Table 1: Distribution of Respondents by Traditional and Social Status 37
Table 2: Distribution of Respondents Regarding Knowledge
   About TMS 37

Chapter 6
Table 2.1: Data Description of the Main Variables 77
Table 4.1: ADF Test Results 82
Table 4.2: Toda-Yamamoto Causality Test Result 83
Table 4.3: Johanson ML Co-integration Test result for FDI and Growth 83

Chapter 8
Table 1: Defence Value for Money Reports 101
Table 2: MOD Contract Types 1980-2010 112

Chapter 9
Table 2.1 Details of respondents' demographic character 134
Table 2.2 Cronbach Alpha Instruments in Each Pillar 134
Table 3.1 level of KM practice and KM pillars 135
Table 3.2 Rank of KM pillars based on the level of problems and priority 136
Table 3.3 Comparison ranking of pillars—academic vs. non-academic staffs 136
Table 3.4 Comparison rankings of desired KM pillars
   academic vs. non-academic staffs 137

Chapter 11
Table 01 weights of criteria and indicators from AHP... 169

Chapter 12
Table 1: Main motivations and challenges of using OSS......... 187

Chapter 16
Table 1: Factors Affecting the Localization of Technology..... 245
Table 2: Factors Affecting the Localization of Technology in Iran 247

Chapter 18
Table 1: Summary of the interviewee involved and projects assessed in the survey..278

Chapter 19
Table 1: TNF and CNF communities...292
Table 2: Combined-Use SUDS Categories Created from Original SUDS Data..296
Table 3. Active permits on the TNF and CNF in CY 2007....................................297

List of Figures

Chapter 2
Figure 1:Cassava Lifter..21
Figure 2:Cassavamanual chipper...................................... 22
Figure 3:Motorized cassava chipper..22

Chapter 3
Figure 1:UNDP'S approach to promoting sustainable livelihoods 30
Figure 2:Variables determining the rate of adoption of innovations..................... 33

Chapter 4
Figure 1: The innovation capacity index..52
Figure 2: Schumpeter's waves accelerate..53
Figure 3: Diagrammatic representation Schumpeterian economic development....53
Appendix 1. Structure of the innovation capacity index..56

Chapter 5
Figure 1: Stakeholders Involved in the Ethiopian PPP......................... 66
Figure 2: A Networking View of Actors in Ethiopian PPP............................ 67

Chapter 6
Figure 3.1: Trend of FDI-to-GDP ratio (fdi/gdp), Real GDP (gdp) and Real GDP per capita (gdppc)... 81

Chapter 7
Figure 1. World Fish Production and Food use consumption 1976-2030.. ... .92
Figure 2. Scenario Matrix..93

Chapter 8
Figure 1: Defence Acquisition Tree Model..102
Figure 2: Procurement Route in VfM Reports...105
Figure 3: Transition Types of VfM Projects...108
Figure 4: Presence of Level 4 Factors in VfM Reports................................110
Figure 5: Contract Type in VfM Reports..110
Figure 6: A Descriptive Model of Defence Projects Time-Cost Progression, based on MPR studies...115
Figure 7: Venn Diagram of IPT Mechanism..120

Chapter 11
Figure 1: Design Of The Research Methodology..164
Figure 2: In total 68 highways were selected as alternatives for the decision modeling...166
Figure 3: Indicators under each criteria...167
Figure 04 Application of Analytic Hierarchy Process (AHP) in highway rehabilitation..168
Figure 01 Top 20 ranked highways prioritized by using: (a) social benefits, (b) economic benefits, (c) administrative benefits, and (d) capital cost.......171
Figure 02 top ten ranked highways of multi-criteria decision making model.......173

Chapter 13
Fig. 1: The Assets Firms govern to create value..194
Fig. 2: IT audit's role in IT risk governance...198
Fig. 3: IT governance structures, processes and relational mechanisms......203
Fig. 4: IT governance Vs organizational performance.................................203
Fig. 5: The proposed research model..204

Chapter 14
Figure 1: Hybrid Pattern Model..211

Chapter 16
Figure 1: The Localization Trend of Chinese Telecommunication Industries..244
Figure 2: The Conceptual Model of Factors Affecting the Localization of Technology in Iran..250

Chapter 18
Figure 1: High-Level Block Diagram of the proposed approach.................281

Chapter 19
Figure 1: The Tongass and Chugach NFs..292
Figure 2: SUDS and TIM permits active in CY2007 by TNF community....298
Figure 3. SUDS permits for Tongass-area communities on a per-1,000- household basis..299
Figure 4. Tongass TIM permittees' community-of-residence......................300
Figure 5.SUDS and TIM permits active in CY 2007 by CNF community....301

Figure 6. SUDS permits for Chugach-area communities on
a per-1,000-household basis. 302
Figure 7. Chugach TIM permittees' community-of-residence...................... 303

ACKNOWLEDGMENTS

We would like to acknowledge the financial support we have received for the publication of this second edited book from the DST/NRF Research Chair on Science, Technology, Innovation for development(STI4D) SARChI in Tshwane University of Technology of South Africa.

ACRONYMS

AAU – Addis Ababa University
AHP – Analytic Hierarchy Process
ANCSA – Alaska Native Claims Settlement Act
ANILCA – Alaskan National Interest Lands Conservation Act
ANT - Actor-Network Theory
AU – African Union
BOO – Build-Own-Operate
BOOT – Build-Own-Operate-Transfer
BOT – Build-Operate-Transfer
BTO – Build-Transfer-Operate
BPR – Business Process Re-engineering
CBs – Commercial Banks
CDN – Content Distribution Network
CFO – Chief Financial Officer
CGM – Cassava Green Spider Mite
CHIME – Centre for Health Informatics and Multi professional Education
CIO – Chief Information Officer
CMP – Cassava Multiplication Programme
CPI – Corruption Perception Index
CSA – Central Statistical Agency
DBFO – Design-Build-Finance-Operate
DFID – Department of Foreign and International Development
EPCo – Electric Power Corporation
EPRDF – Ethiopian People's Revolutionary Democratic Front
FAO – Food and Agriculture Organisation

FDI – Foreign Direct Investment
FGN – Federal Government of Nigeria
FMOH – Federal Ministry of Health
FTLRP – Fast Track Land Reform Program
GCF – Gross Capital Formation
GDP – Gross Domestic Product
GNHI – The Gross National Happiness Index
GNP – Gross National Product
GPL – General Public License
GTP – Growth and Transformation Plan
HDI – Human Development Index
HEIs – Higher Education Institution
HIS – Health Information Systems
HMIS – Health Management Information System
IBAG – Ibrahim Index for African governance
ICI – Innovation capacity index
ICT – Information and Communication Technology
IFAD – International Fund for Agricultural Development
IKMF – Indigenous Knowledge Management Framework
INFRA – Infrastructure Database System
IPO – Initial-Public-Offering
IS – Information Systems
IWI – Inclusive Wealth Index
KM – Knowledge Management
LDCs – Least Developed Countries
LDO – Lease-Develop-Operate
LPL – Linkages to Public Lands
MADM – Multi-Attribute Decision Making
MCDM – Multi-criteria Decision Method
MCIT – The Ministry of Communication and Information Technology
MDG – Millennium Development Goals
MNCs – Multinational Companies
MoE – Ministry of Education
MoFED – Ministry of Finance and Economic Development
MPI – Multidimensional Poverty Index
NCAM – National Centre for Agricultural Mechanisation
NEPAD – The New Partnership for Africa's Development
NFS – National Forest System
NGO – Non-Governmental Organisation
NIS – National innovation system

NPs – National Parks
OFM – Open Framework Middleware
OFAR – On-Farm Adaptive Research
OSS – Open Source Software
PASDEP – Plan for Accelerated and Sustained Development to End Poverty
PPP – Public–Private Partnership
PS – Proprietary Software
R&D – Research and Development
ROT – Rehabilitate-Operate-Transfer
RSDP – Road Sector Development Program
SADC – Southern African Development Community
SEUSISS – Survey of European Universities Skill in ICT of Students and Staff
SL – Sustainable Livelihood
SON – Standards Organisation of Nigeria
SPSS – Statistical Packages for Social Sciences
SSA – Sub-Saharan Africa
SSM – Soft System Methodology
ST – Stakeholder Theory
SUDS – Special Use Data System
TIM – USFS's Timber Information Management Data System
TMS – Tropical Manihot Selections
TNF – Tongass and Chugach National Forests
TVET – Technical and Vocational Education & Training
USFS – U.S. Forest Service
WDR – World Development Report
WFTO – World Fair Trade Organization
WGI – The World Governance Indicators
WSA – Water & Sewerage Authority
WSANs – Wireless Sensor and Actuator Networks

FOREWORD

This second volume for Global Knowledge Exchange Network (GKEN) book publication series is aimed to connect continuously of those learned Africans and friends of Africa knowledge workers (i.e. academics, researchers and practitioners) residing across many countries in order to share their local and global concerns.

The chapters in this book are therefore based on research papers submitted for Third GKEN conference which was hosted under the title of "Exploring the challenge and opportunities for inclusive innovation and sustainable development".

The notable academics and researchers who are active members of the network noted that Africa not only faces daunting innovation and sustainable development challenges but also significant opportunities. There is also a general agreement that suggests that the African development also has the potential to follow a more sustainable path than has been the case in many other parts of the world, where development often resulted in severe environmental problems before the main stakeholders and actors began to get to grips with pollution and natural resource degradation.

The case studies in this volume are only a sample. Many more could have been included but for space constraints. An effort therefore has been made to identify some of the most interesting, promising and pressing issues for this volume. In addition to informing researchers, practitioners and policy makers in Africa, this volume is also meant to reflect the discussions within the wider African and friends of Africa communities about the challenges and responses towards sustainable development and covers the themes of innovation, rural development, good governance and other diverse issues including the appropriateness and application of Information and Communication Technology (ICT) in developing countries.

No continent is more in need for the potential benefits, which science, technology and innovation can bring than the African continent. However, technology cannot yield its potential benefits without appropriate cooperative social initiatives organised by Africans for Africans to provide solutions to all problems. Therefore, devising ways to deliver much greater benefits to all the peoples in Africa is a fundamental need for the continent.

In addition to exploring the potential benefits science, technology and innovation bring, their contributions highlight and assess the scope for scaling up these innovations coming from all sources including indigenous knowledge systems to make an impact and ultimately create an equitable world for all to live in harmonious and sustainable manner.

Amare Desta, Mentesnot Mengesha and Mammo Muchie

INTRODUCTION

We are living in a continually changing environment that touches all aspects of our lives and also generates new challenges along the way. The problems facing humanity have been tackled by fast pacing findings of science and technology during different eras. To respond to our current complex and multilevel needs and challenges it is expected that the persistence of innovative ideas are important. In our globalised world innovations have many routes of transmission and methods of dissemination.

However, the technology divide among the rich and the poor between and within countries will remain a visible barrier that needs close attention. The narrow gap of technological divide and knowledge dissemination should be reduced in order to ensure sustainability and progressive continuity.

Knowledge in its different formats and means of exchange is a tool to engage people in dealing with numerous challenges. On the one hand, the role of knowledge in enriching innovation aiming at improving the wellbeing of humanity is paramount and on the other hand, the amalgamation of innovation and technology moving the limit of knowledge to the next higher level is a daily phenomenon ensuring socio-economic progress. The power of innovations supported by technological advancement has helped to minimise many ills of our society. Nevertheless, developing countries are lagging behind both capturing and also in benefiting from it. This widened divide need to be addressed and a measure to narrow it should be seriously taken into account. Furthermore, the compatibilities of innovation with sustainable development are areas of interest for those working in the developing countries.

Obviously, the support and applications of Information and Communication Technology (ICT) is vital to move development initiatives forward. The question however is how these important ICT applications would be used appropriately, effectively and efficiently.

This book addresses issues of innovations and sustainability on the one hand and the applicability and policies of ICT to the countries mainly and not exclusively in developing countries of the Global South.

Some of the chapters in this volume also focus on the ICT policies issues in a holistic manner at a national level. They clearly explain the infrastructure development and its regulatory arrangements. The ICT policies that stipulate how the different section of the government departments, businesses, industries and other organisations could comply with regulations and policies efficiently would help the smooth delivery of services. In many occasions the country's development will be dependent on the efficiency of the ICT management and arrangement. Therefore, policies that would respond to the demand to public and private services will help the socio-economic growth and development of the country and help us to understand the major problems and opportunities for African innovation systems.

Hence, the book comprises three major parts as thematic emphases and twenty chapters. The first part discusses the dynamics of innovative and sustainable developments from the wider socio-economic perspectives while the second part emphasises on the appropriateness and application of Information and Communication Technology (ICT) in the context of developing nations. The third part focuses on policy related issues to unpack what developing countries face at this moment in time with possible recommendation to signpost positive roadmap.

Kingstone Mujeyi in chapter one argues that Zimbabwe has undergone a substantial industrial reconfiguration in which the informal metal manufacturing sector supplied increased demands of smallholder farmers' agricultural mechanization technologies. The author introduces following implementation of extensive redistributive land reform since 2000, the country plunged into a decade long economic crisis in which formal large scale industrial entities either scaled down operations or shut down altogether, rendering many workers jobless. The author further discusses the collapse of the formal industry resulted in constrained supply of agricultural mechanization technologies to the smallholder farming communities. This came as an opportunity for retrenched workers, some of whom had acquired substantial skills and experience, to feel the void by creating employment for themselves through engaging in informal metal manufacturing of agricultural mechanization technologies for the smallholder farmers who had acquired land under the land reform program. According to the author, since the new farmers had different mechanization needs from the former large scale farmers, the informal entrepreneurs had to develop new innovations to supply appropriate technologies to the farmers. Finally, the author argues despite this crucial service provided by the informal entrepreneurs, these grassroots innovations have not received adequate attention. Using data from a survey of the informal metal manufacturing industry collected in 2012, the author analyses factors that influence innovations development in the informal sector. He concludes that various social, economic, agronomic and environmental factors were highlighted as influencing innovations development informal entrepreneurs as they seek to supply appropriate agricultural mechanization technologies to smallholder farmers and processors.

In chapter two **O. I. Ogunyemi** looks Africa as a continent technologically lagging and yet possesses her own indigenous technology. To substantiate the argument, the author evaluates the activities of the National Centre for Agricultural Mechanisation in Nigeria which is happened to be the only agricultural mechanisation centre in sub-Sahara Africa. The author used descriptive statistics: graphics, pictograms and interview of some staff of the organisation to elicit information that are asymmetry in nature. On the result section the author shows that the organisation has fabricated home grown agricultural machines and equipment for small and medium scale farming using indigenous technology with local material contents. He concludes that with better funding and support from international donors and development partners, the fabricated equipment and machines will be less costly for the farming users. The author also restated that the extension linkages that are currently being done will bear the desired outcomes to improve market access and utilisation of the machines by small farm holders and commercial farmers towards food security.

Edwards Adeseye Alademerin in his part in chapter three introduces that extension models pay greater attention to farmer's knowledge exchange interaction and farmer first approaches unlike the top down and transfer of technology where actions of farmers were mostly receptive in areas of adoption of such imported technologies. The author argues that farmer to farmer knowledge exchange system takes its roots from indigenous knowledge development which is a predominant feature of peasant cassava farmers in Southern Nigeria. The author examines the nature of rural innovation and indigenous knowledge systems development, the theories and models of innovation and indigenous knowledge, techniques of information and knowledge dissemination among cassava co-operative farmers, and the benefits of co-operative Science, Technology & Innovation for Sustainable Future movements in knowledge exchange among others in Southern Nigeria. The author recommends on better ways to harness these potentials in the rural areas for the overall National and continental sustainable agricultural development.

Bedru B. Balana in chapter four introduces innovation in developing countries as an engine to social and economic development which very often overlooked in the development process. The chapter reviews the concept of 'innovation' as a social, economic as well as technical process. The author presented the innovative examples of Ethiopian shoemaker Bethlehem to illustrate the huge untapped potential of innovation for development. The author discusses key barriers to innovation and highlights a way of tackling these barriers to innovation and suggests policy options particularly for developing country governments, including Ethiopia. The policy regimes, the author suggests include to develop pro-innovation policies; allow free flow of information; gear education and research to provide evidence-based; human capacity-building to promote innovation; and promotion of a wider awareness and public understanding of innovation at the practitioner, programme and policy levels to ensure innovation gets effectively onto the Development Agenda.

In chapter five **Temesgen A. Weseni and Richard T. Watson** propose the possibilities and features of partnership efforts to promote the potential of creating a partnership between the public and private sectors of developing economies. The authors argue private firms are more flexible organizations compared to that of public ones. They further debate about the challenge for the public sector to achieve its public servicing goals without harnessing the possibilities of partnerships with the private firms. The authors recognise governments in developing economies are investing a lot to achieve the delivery of public information services, however, they indicate that most of these implemented initiatives have failed and there is a gap in Information Systems (IS) literature regarding the role of PPPs as a mechanism to assure proper delivery of public information services. Finally, the authors reiterate the outcome of their investigation could be used by policy makers and practitioners as a tool to assess the role and success of PPP initiatives in developing economies.

In chapter six **Yesuf M. Awel and Tsehaye Weldegiorgis** investigate the causal link between FDI and economic growth in Ethiopia. Using annual data ranging from 1974-2010 and employing the Toda-Yamamoto (1995) bivariate causality test, the authors find that no causality running from FDI to growth or vice versa. However, the authors identify that there was an evidence of co-integration between FDI and growth. They also suggested that developing countries like Ethiopia should formulate policies attracting FDI in to economic sectors that could harness the benefits of the FDI outweighing the costs of hosting FDI (like profit repatriation to FDI sending economy). They also suggest with further availability of data, future research should examine the causal link in a multivariate framework while addressing the issue of structural break.

Worku Jimma, Daniel Adjei-Boatengand Nicolas Van Vosselen in chapter seven introduce that aquaculture as the fastest-growing food producing sector in the world as the world's human population continues to expand beyond 7 billion, and the reliance on aquaculture products as a cheap source of protein continues to increase. However, they argue over the years various aspects of this sector have come under scrutiny due to the negative impacts on the aquatic environment and more importantly the sustainability of the aquaculture industry itself with respect to environmental issues and the utilisation of resources. The authors et. al suggested the development of scenarios on the future of aquaculture is a paramount if the socio-economic, environmental and ecological impacts of aquaculture are to be ameliorated and more importantly, to focus aquaculture on the path of sustainability. The study creates scenarios on the internal driving forces that could shape the aquaculture practices and direct it on the path of sustainability.

In chapter eight **Ermias Kebede** presents the UK Defence Acquisition System, which aimed to grasp the nature of the UK Ministry of Defence (MOD) engagement with its Industrial partners in the acquisition and delivery of defence equipment projects. The chapter argues that a relational contracting approach, which emphasises and strengthens partnerships for transaction-specific and long-term buyer-supplier relationships, is best placed to deal with transaction costs manifesting in these specific market exchanges. The author suggests the approach to acquisition should be applied to deliver projects, which are highly specific and not generic equipment procurement. The author shows through the use of Transaction Cost Economics theory the link between major defence projects, acquisition policy, and transaction costs have been shown. The prevalence of cost and time increases in major defence projects, as shown in the analysis of National Audit Office reports, is attributed mainly to the poor management of transaction risks and the optimism-bias in defence acquisition. The author explains about optimism-bias in transaction cost terms as the combination of behavioural uncertainty (lack of information regarding possible future states) and opportunism (self-interest seeking, with guile). Finally the chapter concludes in order to remove these negative effects on defence acquisition, which create the transaction costs, a set of recommendations have been presented in the conclusion. Finally, the author undertakes a brief critique of existing theories that have informed policy in Africa in order to define a new research agenda to provide metaphors, heuristics, alternatives and the resources for an emancipatory epistemology and practice in the rebuilding of African integrated national systems of political economy, production and innovation.

Haftamu Ebuy, Rahel Bekele and Worku Jimma in chapter nine present the assessment level of Knowledge Management (KM) practices in Jimma University. The authors consider the four KM pillars, namely technology, leadership, organization and learning to examine the perceptions of staff on the current KM practices and ranking of the four pillars from the most to least problematic and the desired conditions to prioritize among the pillars to improve future KM practices of the University. The chapter aim to determine the existing knowledge environment and provide an input regarding the current status, gap, capability and guidance on improving KM practices by referring the area of the four Knowledge Management pillars. The authors employed both quantitative and qualitative study method to carry out the study and revealed that the perception level of Knowledge Management practices among academic and non-academic staff of the University is different. The authors concluded that technology was least problematic while leadership as the most problematic among the four pillars with respect to the current Knowledge Management practices in the University.

In chapter ten **Melisachew Adane, Lemma Lessa and Solomon Shiferaw** discuss that Information Communication Technologies (ICTs) play substantial role in providing efficiency and effectiveness in different sectors and medical education as well as quality health service delivery. The authors argue however, there is no adequate information on the level of knowledge and utilization patterns of ICTs in the health sector in Ethiopia. The authors aimed at assessing the knowledge, attitude and utilization of ICT among students and health care providers in a teaching Specialized Hospital in Ethiopia. They conducted survey in the hospital, a total of 403 study participants were randomly selected from undergraduates, residents and health care providers based on their proportional population size. The authors identify that students and Health Care Providers have low knowledge level and poor utilization status of ICT for academic purpose and for the needs of service delivery. The authors conclude that the need for improving the existing ICT course in the curriculum to be more skill-oriented and also formal in-service ICT related trainings for the health care providers.

In chapter eleven **Tamirat Fikre Nebiyu** presents the decision on highway investment in Ethiopia involves several stakeholders and many attributes. Hence, Multi-criteria Decision Method (MCDM) becomes vital. He used Analytic Hierarchy Process (AHP) and relative weighting to rank the highways for rehabilitation. The author selected sixty eight major highways covering the entire nation and four core criteria namely: social benefits, economic benefits, administrative importance, and capital cost to work out the roads hierarchy and evaluation models. The author concludes that rehabilitating these regional highways would equally be feasible as those which radiate from Addis Ababa. The author reiterates that the final rank of the highways was highly sensitive to changes in data quality, transformation functions, and utility value assignment approaches.

In chapter twelve **Florida Alemayehu and Lemma Lessa** write that in Ethiopia, higher education institutions are one of the major consumers of ICT. The authors argue given the fact that these institutions have limited resource and budget and on the other hand high demand for information technology, the need to consider the Open Source Software (OSS) alternative to balance the situation as apparent. They further argue it is because open source solution is attracting too many users these days from the developing world mainly because of the economic advantage it offers. Through their investigation at Higher institutions, the authors revealed that more than 90% of the applications at the server level are OSS in 83 % of the target institutions. The authors find at the desktop (workstation) level no action was taken in any of the institutions regarding OSS yet. However, the authors also reveal that attitude towards OSS among ICT officials was found positive. Finally, the authors conclude that wide spread usage of pirated software, lack of awareness, resistance to change and training needs found to be the foremost barriers for the adoption of OSS at the desktop and recommend possible short and long term future actions and directions.

Mengistu Bogale in chapter thirteen identifies Information Technology (IT) governance problems are critical factors for success or failure in organisations that are IT dependent for information provision and business operations. The author discusses that in the developed world, different models of IT governance are developed and the agenda is fully supported by IT auditing process. Nonetheless, in Ethiopia, there are no recorded studies as to how the problem of IT Governance is analysed and resolved the author confirmed. Mengistu uses the field study (survey) research that employs both quantitative and qualitative methods to better substantiate the findings of the study. Mengistu Bogale contributes to the IS knowledge-base by incorporating IT auditing in IT governance, providing recorded evidence of IT governance practices in developing economies such as Ethiopia and that enable to develop an appropriate IT governance framework for Ethiopian context. The author reiterated such a framework may also be used in countries with the same situation as Ethiopia such as those in Sub-Saharan Africa. This will enable businesses to create value from their IT investments.

Asebe Jeware and Nassir Dino in chapter fourteen introduce the concept of implementing software architectural pattern in designing software structure as key for system development process efficiency accepted by the communities of software engineering. The authors and Nassir Dino argue that since the implementation of a single or a uni-architectural pattern throughout in solving system recurrent problems, and designing the whole system has its own limitations when the concept of using more than one architectural pattern emerged. However, on the other hand the authors appreciate that monolithic approach, in general, has a limitation of preserving the advantage of multi-pattern capabilities that is not found fully in one pattern. The authors contributed to a design highly efficient system architecture model known as hybrid architecture that allows considering more than one architectural pattern in a system. And finally, they suggest a model with the procedure in implementation of hybrid architectural design to different applications.

Amare Desta in chapter fifteen discusses despite improvements in access to and use of ICT around the world, there is evidence of a persistent digital divide between and within countries and argues nowhere is the ICT gap more evident than in the education sector. The main stakeholders (i.e. students and educators) in developing countries often lack access to computers and software, and educators are not always trained in how technology can aid in learning according to the author. He argues that there is a huge debate about the role and extent of the information and communication technology ICT in transforming the pedagogical practices – especially in developing countries like Ethiopia. Despite the fact that ICT related infrastructures in most developing nations are limited, the author claims that ICT based technologies have the capacity to transform the provision of all higher education globally. The author conducted his research at two universities in Ethiopia Addis Ababa University (AAU), which is one of the oldest tertiary level education institutions in Africa with a current enrolment of over 40,000 students in its regular and continuing education programs, and Unity University which is the first privately owned institute of higher learning in Ethiopia with current enrolment of over 7,000 students to explore the extent of Information and Communication Technology (ICT) in supporting pedagogical practices in developing countries.

In chapter sixteen **Mohammad-Ali Shafia, Mehdi Mohammadi, Ali-Reza Babakhan, Mohammad Ameri, Faramarz Lotfali** introduce that in today's competitive world, different methods are used for acquisition and development of technology. The authors argue that the ultimate goals for acquisition and transferring technology in developing countries are absorption and localization of technological knowledge. However, due to the interference of some obstacles and factors, the very goals of technology application do not seem to be achieved according to the authors. Like other developing countries, Iran confronting some obstacles in its path towards localization of technology. Mohammad-Ali & et. al identify the relationships of factors with localization of technology at micro and macro (i.e. firm) levels. The author's findings suggest that the companies which are interested in localizing a technology & seeking the success should invest primarily on five influencing areas of learning styles, technological cooperation, technological adaptation, absorption capacity, and main localization activities.

Harmella Ayalew in chapter seventeen discusses the extent of Knowledge Management Practices in Development and Humanitarian Aid Organizations in Ethiopia. The author explores factors that influence the exchange of knowledge in aid and development organizations. The author argues through Knowledge Networks and communities of practice development, organizations can promote partnership, teamwork, with the free exchange of knowledge and best practices between the organization, operational partners and international agencies. The author asserted that while the overall lack of well-defined Knowledge Management strategies is a challenge, there are, however good practices and organizations are rapidly moving towards building on and further developing their knowledge management practices in the coming few years. Finally, the author recommends that the vast majority of the firms examined are doing at least some initiatives in the name of either Information Management and/or Knowledge Management which need to be continually encouraged for further progress.

In chapter eighteen **Mesfin Kifle** introduces challenges and opportunities in organizational requirements development so as to enhance collaborative learning. To achieve this, the author attempted to investigate ideas of collaborative learning to understand organizational context and design a technique that support learning as well as well as identify the challenges and opportunities. Through his qualitative research approach, the author finally concludes that ontology based learning oriented approach for organizational requirements development is a comprehensive approach to explore organizational situation. The author also reiterates ontology based learning is a cyclic approach to make use of organizational ontology to support learning.

Mekbebe Tessema, Robert J. Lilieholm, Dale J. Blahna, Linda E. Kruger, Joanna Endter-Wada in their part in chapter nineteen discuss rural communities in the western U.S. and Alaska as highly dependent upon surrounding publicly-owned forests for various economic and non-economic needs. As a result of limited data that hampered understanding these community-resource linkages, community interests may not be adequately considered in forest management and policy development the authors argued. The authors conduct a study to understand and describe community-resource use linkages surrounding the Chugach and Tongass National Forests, using place-level socioeconomic data from the 2000 U.S. Census in combination with permit data from the USFS's Timber Information Management Data System (TIM) and Special Use Data System (SUDS). The study identifies some limitations of the permit data and its potential use in community impact assessments. Despite these limitations, the methodologies utilized demonstrates how secondary data (TIM and SUDS data, in combination with U.S. Census data), could be used to describe Alaska residents' socioeconomic profiles for communities located in proximity to NFs. The authors argue the approach of studying the profile of resource dependent communities and their linkage to land can also be adopted very well in a country like Ethiopia where more than 85% of the population are resource dependent to support livelihoods, and help to advice policy makers to develop sustainable land and resource management policies.

In chapter twenty **Mentesnot Mengesha and Mammo Muchie** discuss Power and Empowerment in the Context of Public Participation in Urban Development Projects in Sub-Saharan African cities. The authors argue as in many parts of the developing world, notably in sub-Saharan Africa, urbanisation is seen to be a fast growing phenomenon, then to formulate sound urban policy, it is essential to see a range of areas whereby the public have full ownership to its urban development initiatives. Mentesnot and Mammo discuss the examples of contesting theories in areas of power and empowerment in the context of public engagement in urban development projects applicable to sub-Saharan Africa. In the first part their review they discuss power in the context of public engagement. The section discusses Foucault's and Luke's views on power and its relation to public and community participation. While in the second part the authors discuss empowerment and conceptualising it in relation to day to day applications in public and community engagement viewing it from developing African cities. The Paulo Freire's concept of community empowerment is discussed, as a transferable experience that could be applicable in similar situations. The urban development projects in sub-Saran African cities would only be sustainable if and only if the collective ownership of the public is guaranteed. The ownership is possible when the public fully engaged in the process from the outset of projects. Finally the paper contributes to the timely ongoing debates of public and community participation and how power is dispersed and empowerment is secured in sub-Saharan African cities.

By and large, the book brought together different ideas in the wider ICT governance, sustainability and policy from experiences drawn from developing countries including Ethiopia, Iran, Nigeria and Zimbabwe. These experiences have many areas in common that the readers will take to reflect on. We believe, the ICT is a less trodden area when comes to the developing world.

We also believe that this volume will instigate more discussion among academics, research students and practitioners to fill the vast gap of knowledge why developing countries are left behind in the ICT and knowledge management. The book, we strongly believe is a new addition, with variety of perspectives, to enhance our knowledge on this particular field.

Mentesnot Mengesha, Amare Desta and Mammo Muchie

PART I:

INNOVATIVE AND SUSTAINABLE DEVELOPMENTS AS SOCIO-ECONOMIC SOLUTIONS

CHAPTER I:

Grassroots innovations for Zimbabwe's smallholder agricultural transformation: Evidence from informal metal industry survey

Kingstone Mujeyi

1.
INTRODUCTION AND BACKGROUND

Following extensive land redistribution between 2000 and 2010 under the auspices of the fast track land reform program (FTLRP), major changes in Zimbabwe's agrarian economy have occurred that have resulted in significant shifts in agricultural production and the functioning of markets in general (Mujeyi, 2010). The land reform officially benefited 168,671 black indigenous families, comprising mainly the rural poor across 9.2 million hectares (Moyo, 2011). A severe economic meltdown coincided with this period, in which the informal metal agribusiness industry supplied increased demands for agricultural mechanization inputs of the broadened farming clientele base. While Zimbabwe had a highly developed and diversified agricultural engineering, mechanization and irrigation supply industry before the FTLRP era, government and private sector efforts towards developing appropriate small scale machinery, implements and tools to respond to the needs of smallholder farmers remained minimal (Nazare, 2005).

During the economic crisis period, the formal industry faced critical constraints in accessing foreign currency to import raw materials, components/spare parts and machinery, resulting in constrained supply of agricultural mechanization inputs to farmers and other users. Many employees, some with vast experience, were laid off from formal employment during this period as the formal industry scaled down operations or closed altogether. This gave rise to an influx of small scale informal operators into the metal manufacturing agribusiness supply chain as a crisis-induced employment strategy (necessity entrepreneurship) but also as an opportunity and innovation-induced entrepreneurship.

Their operations, which are mainly undertaken in urban centres and rural growth points in proximity to the farming communities, are driven by existence of unexploited opportunity for innovation and invention of appropriate smallholder farming equipment to cater for increased demand from the broadened farming clientele base owing to the redistributive FTLRP. The operators in this industry rely on locally available resources such as scrap metal for raw materials and other less expensive factors of production.

The rise and intensification of innovative informal metal industrial activity during the economic meltdown period was expected to slow down and eventually disappear with economic recovery. However, since the adoption of the multiple currency system in 2009, which brought the country's economy back on the growth and recovery path, the informal economy, particularly the metal agribusiness sector has not shown any signs of slowing down. The informal metal industry and the informal economy in general, have not only withstood the test of time but have actually exploded during the past decade to become the leading employer.

Scholarly work on metal technological innovations in Africa has put more emphasis on the expansion of scientific knowledge in metallurgy and mechanics within formal enterprises and universities (Manyati, 2013). Research has not cascaded down to tap into innovations occurring among indigenous people within the informal sector. As such, an intellectual gap exists with regard to the micro-processes and interaction which occur within the informal metal manufacturing industry in the production of innovative technologies. The indigenous knowledge systems and skill within the informal sector which propel technological innovations remain under-researched. Innovations in Africa have largely been found to focus on and be associated with discoveries made in universities and/or research centres (Manyati,

2013). This situation is highly problematic in that the realities in Africa are neglected. These require the development of appropriate and home grown technologies as the solution to developmental challenges such as food insecurity, poverty, environmental sustainability and diseases (ATPS, 2010). For instance, food insecurity problems and low productivity in African agriculture have in part been aggravated by the blind adoption of technologies that are inappropriate to African agricultural practices.

Throughout Africa, the documentation of innovation by African people has not yet received the attention that it deserves. Existing literature has deliberately focused on the interactions which occur at a macro level among universities, agricultural researchers, science granting councils, government agencies and commercial farmers (Manyati, 2013). The dynamic linkages and social interaction which occur at a micro level are currently under-researched and they differ from the interaction observed in formal industries and firms by various authors. In addition limited research has been carried out to determine the drivers of informal metal manufacturing firms and entrepreneurs to develop innovative technological designs.

The development and scaling up of locally grounded technological innovations could ensure the development of a sound industrial and technological base which would address local problems of food insecurity and low agricultural productivity. This could enable the continent to emancipate itself from the dependence on and appendage to the global economy. This study uses baseline survey data collected between August and September 2012 to highlight innovative production and marketing channels being used by the informal metal agribusiness entrepreneurs in helping transform smallholder agriculture by supplying agricultural mechanization input requirements of farmers in rural Zimbabwe. It finds the innovators to be fairly young and highly educated and skilled entrepreneurs producing exciting innovations for appropriate mechanization of smallholder agriculture.

2.
EMPIRICAL EVIDENCE OF GRASSROOTS INNOVATIONS

Manyati (2013) asserts that inconclusive contestations and debate have been generated among scholars regarding the ability of grassroots people in developing countries to innovate new products. Indigenous innovations have been acknowledged in academic and developmental literature as key in facilitating transformation of developing and transitional economies to full industrialisation (see ATPSN, 2010). Following the industrial reconfiguration (disintegration and informalisation) of the Zimbabwean economy, several studies have been undertaken to document changes in the organisation and the products produced in the informal metal industry (see Sachikonye, 2006; Luebker, 2009; Chirisa, 2009; Makochekanwa, 2010; Simpson, 2010; Mawomo, 2013, Makate, 2013; Manyika, 2013 and Manyati, 2013). However, not much has been done to document technological innovations developed in the informal metal industry. Manyati (2013) only went as far as illustrating the knowledge and skills of informal metal manufacturers which enable them to develop innovations. This study sought to identify what motivates the entrepreneurs to innovate and to determine the extent of their innovativeness and how the innovations are meeting the needs of smallholder agriculture in Zimbabwe.

Innovation in metalwork has historically occurred throughout Zimbabwe and Africa through such activities as iron working and blacksmithing that have existed at least two centuries before any form of European contact was established (Mhone, 1996; Wakandigara, 1999; Mararike, 2001; Mushipe, 2007; ATPS, 2010; Manyati, 2013). Ancient metalworking, according to Thandlana (2005), involved the use of indigenous technology to fabricate such metals as iron, copper, gold, tin, copper and copper alloys such as bronze and brass into various tools and weapons. Substantial increases in indigenous population required more efficient tools (Otsuka and Kdirajan, 2005). As such, innovations in metal working were chiefly driven by fundamental transformations in economies and societies as a result of immigration patterns and cross-cultural exchanges among various African tribes (Garlake 1982). For instance, the Bantu speaking people who entered Eastern and Southern Africa replaced the hunter-gatherer Stone Age culture with metal technology. According to Manyati (2013), the apex of skilled craftsmanship and indigenous innovation development in agricultural implements in Southern Africa can be traced back to the arrival of migrant groups to Zimbabwe plateau.

Technology development and innovation in metal manufacturing in Zimbabwe has been a notable feature since the pre-colonial period (Mararike, 2001; Wakandigara, 1998). The coming in of colonial rule decimated the capacity of indigenous manufacturers in Zimbabwe. The ancient iron smelting industry was prohibited by the colonial government for humanitarian and political reasons. The metal manufacturing activities among other local innovations in farming and agriculture tools were denigrated by settler capital as "inadequate", "backward" and "inefficient" and the activities were thoroughly regulated by the colonial institutions such as the police (Manyati, 2013).

The transformation of the educational system in Zimbabwe significantly contributed to the negation and displacement of the African indigenous knowledge systems (Mararike, 2001; Sagasti, 2003). Imported knowledge and the establishment of formal educational systems through vocational training centres weakened indigenous innovations and technologies (Manyati, 2013). Settler capital during the pre-independence era promoted vibrant research based industrial development policy which facilitated the development of firm level innovations. At that time, Rhodesia (now Zimbabwe) achieved tremendous heights in vocational skills training and technological innovations which saw Zimbabwe becoming the second most industrialised economy on the African continent (Sadomba, 2009).

The first decade after independence in Zimbabwe (1980 to 1990) saw the relaxation of colonial laws restricting the informal sector economy resulting in slight increase in the informal sector activities (Mhone, 1996; Mupedziswa, 1991; Gwisai 2006). The informal economy became a visibly significant phenomenon after the implementation of the IMF- led Economic Structural Adjustment Programme (ESAP). By 2001, the informal sector economy employed 61% of the labour force (Luebker, 2009).

Since the setting in of the economic meltdown beginning 2000 which led to the closure and downsizing of many large-scale and formal manufacturing firms triggered a dramatic upsurge of the informal economic activity to an astronomic employment level of 87.8% by 2008 (Makochekanwa, 2010; Murisa 2010; Gumbo 2013). The metal manufacturing industry has constituted a highly significant proportion of the informal economic activity. From the above information, it can be noted that the indigenous metalwork has existed in varying social contexts since the pre-colonial period up to present. A review of existing literature on technological innovations in the informal metal manufacturing industry reveals that technological innovations in the informal sector in Zimbabwe are scantly documented, hence the need such studies as this one.

3.
METHODOLOGY

In order to fully explore the innovations which originate from the informal metal industry in the design and manufacture of agricultural mechanization technologies (implements, equipment and machinery), this study triangulated qualitative and quantitative techniques in the collection of primary and secondary data in order to obtain a holistic picture of informal sector innovations. The structured questionnaire was used to generate information which can be generalised specifically among agricultural mechanization innovators across Zimbabwe. Qualitative techniques were used to solicit in-depth information from the participants pertaining the social networks and interaction which occur in the process of innovation development.

Data was collected from 602 randomly selected informal operators in the metal manufacturing industry across 15 cluster-sampled districts in 8 of the country's 10 provinces. For qualitative information, purposive sampling was used to target individuals with rich in-depth knowledge and information on technological innovations development processes in the informal metal manufacturing industry. The Statistical Package for Social Sciences (SPSS) was used to capture and analyse the collected data. Seven in-depth interviews were conducted with key informants who design and manufacture smallholder agricultural mechanization technologies. Observation was also used as a means of collecting qualitative data to gain more information with respect to their innovative ways of doing things and also to physically inspect their production processes and product quality.

4.
STUDY FINDINGS

The study finds that besides replication of the formal sector production processes, owing mainly to the absorption of retrenched workers from that sector, the informal economy has also undertaken to improve on existing agricultural mechanization technologies and sometimes invent new technologies altogether, to suit the needs of smallholder farmers and other users such as processors. Following the downscaling of operations by large scale formal firms during the economic crisis period, informal operators moved in to fill the void in supply of appropriate agricultural mechanization technologies to the broadened farming clientele base following the extensive land and agrarian reforms implemented from 2000. These informal operators are mainly comprised of sole proprietors or micro and small enterprises, relying on family, and to a limited extent, hired labour for carrying out business operations.

They are usually not properly registered resulting in them operating on the fringes of the law. Their products are favoured by customers owing to their giving room for product price negotiation and modification of product to suit customer needs. Contrary to widespread view and expectation that entrepreneurs in the informal economy should be unskilled, semi-skilled and uneducated, in Zimbabwe's informal metal manufacturing industry, majority of them have higher educational attainment (secondary education on average) with college and university graduates and post graduates also being found.

4.1 Factors Influencing Innovation Developments in the Informal Sector

Although the average age of active players in this sector is 35 years, it is skewed towards the youthful age group of the 20s indicating the important role that the sector is playing in absorbing unemployed youths. In terms of education, the industry is dominated by secondary school graduates (67%) with over 80% of the entrepreneurs having at least completed secondary education. Those who failed to attain at least primary education are almost insignificant (2.8%). Beyond secondary education the industry has attracted mainly those inclined towards practical vocational training (10.4%) as opposed to the academics (2.2%) as shown in Table 1 below.

Table 1: Educational Level Attainment

Educational Level	Frequency	Percent (%)
Never been to school	9	1.5
Primary School Dropout	8	1.3
Primary School	42	7.1
Secondary School Dropout	57	9.6
Secondary School	398	67.0
Academic College	13	2.2
Technical College	62	10.4
Undergraduate	4	.7
Post Graduate	1	.2
Total	**594**	**100.0**

The number of years of formal education averaging 11 years implies that the informal sector metal manufacturers possess the necessary education which aids them in developing innovations. Majority (81.9%) of the entrepreneurs indicated that they design new innovations. Designing new innovations is done as a way of meeting customer specifications and needs but also a way of improving on efficiency of technology. Innovation in the design and manufacture of agricultural mechanization technologies is motivated by the need to improve on the efficiency of the technology in question. An interview conducted by Manyati (2013), an informal agricultural equipment manufacturer indicated that:

[The reason why we consider our crop thresher to be a unique innovation is because we improve on the existing one and then we come up with a more effective crop thresher which works in a much better way].

The above quotation therefore illustrates the importance of efficiency in the manufacture of agricultural mechanization technologies. The manufacturer even bragged that farmers preferred their newly designed equipment to those manufactured in the formal sector due to its improved efficiency.

Modification is also done as a way of reducing the costs of producing the agricultural mechanization technologies so as to make them affordable to customers. For instance, one manufacturer was observed telling a customer that the price of a multi-crop thresher could be reduced if the customer provides the specifications of the equipment that he could afford. The other option was for the customer to provide raw materials required for the manufacture of the crop thresher which would entitle the customer to paying the labour costs only. Lack of access to credit and other financial services was highlighted as a major constraint to innovations development in the informal sector. Most innovations are motivated by the need to solve a specific problem. For instance, another manufacturer interviewed indicated that he was in the process of designing a combined thresher and grinding mill which would be far much cheaper compared to buying the two separately. This was said to be a response to a problem of lack of crop threshers in the area where the entrepreneurs' rural home area.

Some innovations are driven by the need to respond to changing agronomic and environmental factors/patterns. As farmers increasingly adopt conservation agriculture, the innovators quickly adjust their innovations to come up with appropriate implements ideal for conservation tillage such as direct seeders, jab planters and basin diggers. One innovator indicated that they had to change painting their crop processing equipment from yellow to green after discovering that yellow attract bees more than green. The type of crop grown and commodity produced were found to significantly influence innovation decisions by the metal manufacturers but indications were that there innovators are always on the lookout for technologies that serve multi-purposes for the customers. This was said to give the innovators a competitive edge over their rivals in the formal sector who were said to be slow in responding to customer needs.

The boom in smallholder tobacco production in Zimbabwe was pointed out by the informal metal manufacturers as one that was significantly aided by innovations in the informal sector. Today, most smallholder farmers procure their tobacco handling and curing equipment from the informal sector manufacturers who supply affordable technologies. Such equipment as flue pipes, curt machines, clips, etc. for tobacco curing which used to be unaffordable to smallholder farmers just because the formal manufacturers had designs suitable for large scale commercial farming are now being supplied from the informal sector in appropriate designs and specifications for smallholders at affordable prices. The intensification of informal metal manufacturing activity has given rise to increased response to the needs of rural agro-based households by manufacturing mechanization technologies that use alternative and renewable energy sources from conventional electricity such as solar, manual, gravity and wind. Increasing dependency on renewable energy is critical for ensuring sustainable growth and promotion of a green economy.

4.2. Sources of skills Training for innovation

Skilled artisans who are formally trained in vocational training centres are instrumental in the development of innovations. Manyati (2013) found that trained manufacturers with formal sector experience are principally responsible for the development of innovations whilst non trained manufacturers rely mostly on modifications and copying designs from other sources. However, the study found that majority of entrepreneurs (70.3%) relies on other operators in the informal sector for acquisition of skills for innovations implying that social capital and networks are key for skills sharing.

Table 2: Source of skills training

Skills Source	Frequency	Percent
Government VTCs	96	18.2
Private VTCs	34	6.4
NGOs	1	.2
Other Informal Operators	371	70.3
Other	26	4.9
Total	**528**	**100.0**

Thus, vocational training and formal sector work experience are pivotal for impartation of skills for innovations on other informal manufacturers who are not formally trained. Previous studies also concur with this finding that skills in the informal sector economy are largely derived from the formal sector economy (Daniels, 2010; Manyika, 2012; Manyati, 2013).

5.0
CONCLUSIONS AND RECOMMENDATIONS

5.1 Conclusions

This study establishes that technological innovations in the design and manufacture of appropriate agricultural mechanization technologies occur within the informal sector of Zimbabwe. Social, economic, environmental and agronomic factors combine to influence technological improvements by operators in the informal metal manufacturing industry. Skills development and knowledge transfers also occur within the informal sector. The source of knowledge and skills for innovation is derived through vocational training, on job experience, customer prescriptions and prior experience of the manufacturers in the formal sector. The study also establishes that innovation in the informal metal manufacturing industry of Zimbabwe is a product of social networks involving close acquaintance relationships with both customers (smallholder farmers and other processors) and other informal sector manufacturers. Knowledge and skills for innovations are readily shared among the informal sector entrepreneurs but mainly emanate from experienced operators with formal sector employment experience and formal training from vocational training and technical colleges.

5.2 Recommendations

There is need to develop good understanding of consumer preferences and perceptions of the smallholder farmers or customers of the agricultural technological innovations produced in the informal sector. Customers are key in influencing technological innovations as they demand appropriate technologies that solve their problems. Public policy need to recognise the contribution of the informal metal manufacturing industry and fully integrate its operations into the mainstream economy since it is contributing significantly towards meeting the agricultural mechanization needs of smallholder farming sector.

In order to enhance their business opportunities, there is need to closely look at the issue of protection of intellectual property rights since currently there is no guarantee that an innovator will fully enjoy the benefits of innovation developments when there is no regulatory framework in place. Again, social and cultural factors which facilitate or hinder adoption of technological innovations in rural communities need to be fully understood and shared with the informal entrepreneurs for them to be able to respond accordingly.

Innovation platforms such as national institutions, universities, private institutions and society need to fully promote the technological innovations developed in the informal sector. The success of locally developed innovations from the informal sector depends on the magnitude of support they receive from these institutions. Collaborative work with professional institutions needs to be fostered for the informal innovators to significantly benefit from their innovations.

REFERENCES

ATPSN, (2010). The African Manifesto for Science, Technology and Innovation. Africa Technology Policy Studies. Nairobi.

Chirisa I. (2009). The geography of informal sector operations (ISOs): A perspective of urban Zimbabwe. *Journal of Geography and Regional Planning,* 2(4), 066-079.

Daniels, L. (1993). Changes in the Small Scale Enterprises Sector from 1991 to 1993: Results of a Second Nationwide Survey in Zimbabwe. Gemini Technical Report, US Agency for International Development.

Garlake, P.S. (1982). Pre- History and Ideology in Zimbabwe. *Journal of the International African Institute,* Vol.52(3).-1-19.

Gumbo, T (2013). On Ideology Change and Spatial and Structural Linkages between formal and informal Economic Sectors in Zimbabwean Cities (1981-2010). DPhil Thesis, Stellenbosch University, South Africa. Stellenbosch

Gwisai, M. (2006). Labour and Employment Law in Zimbabwe: Relations of work under neo colonial capitalism. Zimbabwe Labour Centre and Institute of Commercial Law Publications, Harare.

Luebker, M. (2008). Employment, unemployment and informality in Zimbabwe. Concepts and Data for coherent policy making, Issues Paper No 90, ILO Sub regional office for Southern Africa (SRO- Harare).

Makate, C. (2013). Structure-conduct and performance of the informal metal manufacturing industry in Zimbabwe: implications to stakeholders in the agricultural sector. MSc Thesis. University of Zimbabwe.

Makochekanwa, A. (2010). Estimating the size and trends of the second economy in Zimbabwe. MPRA Paper No. 37807. [Available at]: http://mpra.ub.uni-muenchen.de/37807/

Manyati, T. K. (2013). Exploring Grassroots Technological Innovation: A Study of Informal Sector Enterprises Manufacturing Multi-crop Threshers in Harare. MSc Thesis. University of Zimbabwe.

Manyika, O. (2013). Informal Sector as a Source of Employment: A Case Study of Magaba Metal Manufacturers in Mbare. MSc. Thesis. University of Zimbabwe.

Mararike, C. (2001). African Heritage: Our Rallying point. The Case of Zimbabwe's Land reform, Harare, Best Practices Books.

Mawomo, F (2013). Relevance of Socio-Economic Factors in determining the Performance of Metal Manufacturing enterprises in Zimbabwe: Implications of current entrepreneurial performance and agricultural development prospects. MSc Thesis, University of Zimbabwe.

McPherson, M. (1998). Zimbabwe: A Third National Survey of Micro and Small Enterprises, Final Report, Gemini Technical Report, US Agency for International Development.

Mhone G.C. Z. (1996). The Impact of Structural Adjustment on the Urban Informal Sector in Zimbabwe. Issues in Development Discussion Paper 2. Southern African Regional Institute for Policy Studies. Harare, Zimbabwe.

Moyo S. (2011). Three decades of agrarian reform in Zimbabwe, *Journal of Peasant Studies,* 38(3), 493-531.

Mujeyi K. (2010). Emerging agricultural markets and marketing channels within newly resettled areas of Zimbabwe. Livelihoods After Land Reform Working Paper Series. Working Paper 1.

Mupedziswa, R. and Gumbo, P. (1998). Structural Adjustment and Women Informal Sector Trends in Harare, Nordic Africa Institute. Uppsala.

Murisa, T. (2010). Social Development in Zimbabwe: A Paper prepared for the Development foundation of Zimbabwe.

Mushipe, Z.J. (2007). The Informal Sector in Zimbabwe in F. Maphosa (Ed) Zimbabwe's Development Experiences since 1980: Challenges and Prospects for the Future. OSSERA Publications, Harare.

Nazare R.M. (2005). The Status of Agricultural Mechanization Input Supply Chains in Zimbabwe and Strategies for Improving Input Supply. Assessment Mission Report prepared for the Department of Agricultural Engineering and Technical Services and the Food and Agricultural Organization of the United Nations (FAO).

Otsuka, K. and Kdirajan, K. (2005). An Explanation of a Green Revolution in Sub Saharan Africa. *Journal of Agricultural and Development Economics*, 2(1), 1-6.

Sachikonye, L. (2006). The Impact of Operation Murambatsvina/ Clean up on the Working People in Zimbabwe. Labour and Economic Research Institute of Zimbabwe Publications, Harare.

Sadomba W.Z. (2009). Nipping Bottom-Up Industrialisation in the Bud: The State's Folly of Reversing Zimbabwe's Miracle. Submitted to the *Journal of Development Studies*, 2009.

Sagasti, F. (2003). The Sisyphus Challenge: Knowledge, Innovation and human condition in the 21st Century, Limo, Foro Nacicial

Simpson M. (2010). The Informal Economy, SMEs and the 'Missing Middle' in Zimbabwe: Some Observations. Working Paper 9. Comprehensive Economic Recovery in Zimbabwe Working Paper Series. United Nations Development Programme.

Thandlana, T.P (2005). Style, Space and Time: A Critical Analysis of the Chronology and Spatial Distribution of Copper and Copper Alloy Beads from Zimbabwean Iron Age Sites. University of Zimbabwe: MSc Thesis.

Wakandigara, A (1998). Pre- Colonial Copper Smelting, *Zimbabwe Chemist*, 2(2),2-3.

CHAPTER 2:

Indigenous Agricultural Technology in Nigeria: case study of National Centre for Agricultural Mechanisation

O. I. Ogunyemi & A. S. Adedokun

1
INTRODUCTION

Background statement

Africa is often seen and referred to as a technologically lagging continent. This is so because the continent has consistently failed to showcase her home-grown technical prowess in many areas of life. The continent for a long time has been conspicuously missing on the chronicles of technology in spite of the fact that some of the so called inventions have their roots in ancient civilization in which Egypt and Ethiopia are to be reckon with. For example, Egypt contributed to ancient writing technology while Ethiopia has its indigenous method of measuring time before the advent of time measuring instruments of the Arabs, Barbarians, Jews and Romans among others.

Another supporting evidence is the visit of Queen of Sheba (conventionally believed to be located in Ethiopia) to King Solomon to ask him some questions and observed his wisdom during which she gave him gold and spices (Bible: I king Chapter 10:1-13). Gold is an end product of ancient mining technology in Africa and spices talks of the continent agricultural technology of turning green plants, flowers and animal fats into ancient perfume widely acknowledged as important export products from Africa as at that time. In the progression of time, little is heard of Africa technology globally and the sudden disappearance of Africa indigenous technology relegated the continent to the background and placed other continents ahead in global technology incubation.

Statement of the problem

Africa is a food deficit continent. This is primarily due to the type of agricultural practices that are popularly in used in the continent predominantly for subsistence purpose. With subsistence techniques characterised by manual operations, the farming household cannot produce sufficient amount of food required for the economy since it can barely feeds its household members. For the continent to win the battle against hunger and poverty in the post millennium development goals period, it must not only depend on the north for her agricultural technology but as a matter of necessity, develop her home grown agricultural technology which her large illiterate farmers can easily relate with for the production of food for consumption and export markets. In this connection, this paper provides answer to the question: what is Africa effort at achieving home grown mechanisation for sufficient food production?

Objective of the paper

The primary purpose of this paper is to highlights one of Africa's indigenous technology efforts in agricultural mechanisation through the exposition of the National Centre for Agricultural Mechanisation (NCAM) located in Nigeria.

2
LITERATURE REVIEW

Agricultural mechanisation in Africa is still under-developed with small number of industries that manufacture farm tools and equipment (FAO,

2008). Efforts by many countries in Africa towards mechanised agriculture have continued but yielded few results in negligible number countries which perhaps include South Africa and Zimbabwe. This observed mechanised agriculture in Africa is foreign sourced and did not emanate from indigenous technology and therefore often without local content.

Following various development theories like the Rostow theory of growth, for agricultural mechanisation to be achieved, the transformation would pass through some phases. Ruttan (2013) reports five growth stages in line with Rostow theory as traditional society, the pre-conditions for take-off into self-sustaining growth, take off, drive to maturity and age of high mass consumption. In this wise, Africa agricultural mechanisation can be said to be predominantly in the traditional society stage. The drive towards mechanised agriculture involving the use of machine to accomplish a task or any operation in agricultural production (Odigboh, 2000) has continued but without indigenous technology in most African countries.

But according to Ademiuyi et. al. (2013), agricultural modernisation efforts for mass production and consumption by Nigeria has been on through the introduction and development of need-based, home grown agricultural mechanisation technologies. On this basis National Centre for Agricultural Mechanisation (NCAM) was founded in 1974 in Ilorin, Kwara State, Nigeria and no such institution is available in any other African countries since then. NCAM is the only agricultural mechanisation centre in the whole of Sub-Sahara Africa, the other one is in Egypt, North Africa which has not seen the light of the day. The broad objective of NCAM is for the acceleration of mechanisation in the agricultural sector of the Nigerian economy in order to increase the quantity and quality of agricultural products through adaptive and innovative research; extension and commercialisation of proven technologies; organisations of training workshops and seminars; and networking and collaboration with similar institutions within and outside the country. The specific objectives are to encourage and engage in adaptive and innovative research towards the development of indigenous machines for farming and processing techniques; design and develop simple and low-cost equipment which can be manufactured with local materials, skills and facilities; standardize and certify in collaboration with the Standards Organisation of Nigeria (SON), agricultural machines, equipment and engineering practices in use in Nigeria; bring into focus mechanical technologies and equipment developed by various institutions, agencies or bodies and evaluate their suitability for adoption; assist in the commercialisation of proven machines, equipment, tools and techniques; disseminate information on methods and programmes for achieving speedy agricultural mechanisation; provide training facilities by organising course and seminars specially designed to ensure sufficient trained manpower for appropriate mechanisation; promote cooperation in agricultural mechanisation with similar institutions in and outside Nigeria and the international bodies, connected with agricultural mechanisation (NCAM, 2013). The centre is renowned for having the highest assembly of agricultural engineers in sub-Sahara Africa (Ademiuyi et. al., 2013). The centre has developed technologies that have encouraged the use of mechanised tools and technology through the fabrication of a number of machines like manual seed planter, manual seed and fertilizer broadcaster, cassava lifter, peeler, grater and groundnut digger among others with the training of numerous fabricators on how to produce these machines with local content (Azogu, 2009).

Theoretical framework

Agricultural development is an integral part of economic development since it is the desire of every country to achieve food production sufficiency. There are several theories including Rostow stage theory and Neo-colonial Dependence Model that explain economic development. According to Rostow stage theory, development evolves through five stages of as mentioned earlier. The traditional society is a stage that describes the agrarian economy that depends on the subsistence system of farming. This stage is essential for any economy that will move to industrial age but food productivity and sufficiency is assumed away from a country that is classified to be under this development stage.

Most African countries have not been able to leave this stage essentially because of the low level of technical progress in the continent. It is imperative for the continent to strive towards industrialisation through the mechanisation of her agricultural practices. With the exception of South Africa, and a few other Southern African Development Community (SADC) countries, the level of technological penetration in the sub-Sahara African countries is quite low.

The neo-colonial dependence model qualifies development as the menace of underdevelopment emanating from highly polarised capital between the developed and underdeveloped countries. The rich nations of the world have developed their agricultural sector with both heavy and light agricultural machines and equipment that guarantee food sufficiency and security and may not be so concern with what happens to agricultural technology in Least Developed Countries (LDCs). If LDCs agricultural technology improves, it will reduce. This is further buttress by the false paradigm which opined that underdevelopment is as a result of faulty and inappropriate advice given by international experts through some well-defined international institutions which the LDCs belong.

3
METHODOLOGY

The study adopted interview of some staff of NCAM and secondary sources of information. The information was analysed using descriptive statistics: pictograms and graphics and qualitative presentation.

4
RESULTS AND DISCUSSION

Level of production: The centre produces on commercial level and sells machines and equipment to local and foreign farming users. NCAM has fabricated over 5000 agricultural machines and equipment since its inception in 1974. The local content of the machines varies from 75 to 100 per cent. Three of such equipment are shown in Figures 1 to 3

Figure 1: Cassava Lifter

Figure 2: Cassava manual chipper

Figure 3: Motorized cassava chipper

Figure 1 is NCAM Cassava Lifter which is a device for uprooting cassava tubers. It has a frame to which a foot-board and an immovable gripping-jaw are attached, a handle to which the frame is hinged. The capacity of the device is 2000 plants per man-hour. Figure 2 is Cassava manual chipper with capacity to process 350kg of cassava per hour. Figure 3 is a Motorized cassava chipper developed by NCAM. It has 1.2 tons capacity per hour. The machine may use either automotive gas oil (AGO) or premium motor spirit (PMS). Other selected machines developed by the centre are:

Hand seed planter: a device for planting seeds like maize, soya-beans, guinea-corn consisting of a seed funnel, seed tube, handle, jaw-type of soil opener and seed spacing adjustment. It is designed to drop one or two seeds from the seed funnels at a time.

Vegetable slicer: This slicer is used in slicing carrot, okra, tomatoes and other vegetable crops especially for small scale operators. It is of 30 kg/hr, 20 kg/hr and 15 kg/hr for tomatoes, okra and carrot respectively in terms of capacity. It consists of presser tray, frame and cutting blades.

Tractor mounted groundnut digger: This is a tractor mounted implement for uprooting groundnut plants that are harvest ready. This farm implement can be used on any mechanised groundnut farm to reduce manual labour during harvesting. It is made up of cutting blades that help in cutting the roots of the groundnut vines and loosens the soil. It has a mean capacity of 0.53 ha per hour.

Multi-purpose thresher: It is used for threshing cowpea, guinea-corn, maize, rice and other grain crops. The features are hopper, the threshing unit consisting of beaters welded onto cylinders and the concave. For cowpea, it has a capacity of 200 kg per hour and good for medium-large scale entrepreneurs.

Triketor- This is a 3-wheeled mini-tractor completed and launched in August 2013 by NCAM. The centre collaborated with Bespoke Design Concept. The tractor is reputed to be the 'Made in Nigeria' tractor with almost 100 per cent local content. All the materials used for making the tractor were obtained locally.

Mode of dissemination of information to farmers: The centre gets the farmers to know of its products through its extension staff and seminars. Other means are print media, workshop, conference and training.

Problems of the centre: The identified problems of the centre include gross underfunding, inadequate training for staff which is related to paucity of fund, inadequate power supply and problems relating to bureaucratic set up. The underfunding is connected with low level of investment in research and development (R & D) in Africa generally compared with western countries. The centre also enjoys little or no international funding as subvention from the Nigeria Government is the major source of fund.

Reducing imported agricultural machines: NCAM is in a position to reduce import bills on agricultural machineries through the list of all the machines it has fabricated with the materials used mostly sourced locally. There is no agricultural machineries that enter Nigeria without the approval of the centre for the purpose of ensuring that the advancement of local and indigenous agricultural technology is not jeopardised. The design of some equipment for agricultural purposes in the west sometimes may not have the ecological composition of Sub-Sahara Africa countries in contemplation. This definitely may affect the usefulness or otherwise of such agricultural equipment. This is the reason why the Federal Government of Nigeria made it mandatory to all importers of agricultural machinery and equipment to get approval from the National Centre for Agricultural Mechanisation before such can be allowed into the country.

NCAM as potential hub of Africa agricultural mechanisation: The centre is achieving its objectives and it is considered as a potential centre for the development of home grown agricultural machineries in Africa. The centre as the only one in Africa stands to accommodate interested Africans for training and workshops with facilities for lecture deliveries and accommodation. The post millennium development goals era which indicates a challenge for Africa to fast-track her pursuit of mass production and consumption to reduce the dependence of the continent on the West underscore the importance of NCAM. The short-coming of neo-colonial dependence growth model also indicates the importance of home grown mechanised agriculture in Africa for the continent to pass through the stages of development. It is time for agricultural technology renaissance in the Africa continent and this could be led by Ethiopia, Egypt, Nigeria and South Africa because of their geographic strategic positions in the continent. Ethiopia is the only country in Africa that was never colonised and she is the symbol of the continent strength and Africanism. Egypt is a country with rich history and a major contributor to Africa indigenous technology while South Africa, is a country with history of long stay of colonisation and apartheid regime and Nigeria, the most populous black nation in the world. The first three largest economies in Africa- South Africa, Egypt and Nigeria are among the four mentioned countries. All these countries could serve as the centres of indigenous agricultural technology for their respective sub-regions.

5.
CONCLUSION

The development of indigenous agricultural technology in Africa is on through NCAM, the only agricultural mechanisation centre in Africa. NCAM has developed many machines of local contents varying from 75 to 100 per cent is meeting up with its objectives. The centre has the potential to improve its activities with support and funding from international donors and development partners. With enhanced local and foreign funding the fabricated equipment and machines will be less costly for farmer and the extension linkages that are being done will bear the desired outcomes. This has the inherent advantage of improved market access and utilisation of the machines by small farm holders and commercial farmers towards sufficient food production and security. The study equally recommends the establishment of indigenous agricultural mechanisation centre in other sub-regions of Africa: North, South and East Africa with NCAM as reference. Establishing Agricultural technology incubation centres in Africa will promote exchange of indigenous knowledge required in building formidable agricultural equipment necessary to drive the intended agricultural development of the continent.

REFERENCES

Ademiuyi, Y. S, Azogu. I. I, Dada-Joel, O. T and T. Wakatsuki. 2013. Nigerian policy on agricultural mechanisation and lowland development: serif achievement strategy. www.kinki-ecotech.jp/.../AZOGU%20et%20al-WorkshopPaper.pdf, [Retrieved September 20, 2013].

Azogu, I. I. 2009. Promoting appropriate mechanisation technologies for improved agricultural productivity in Nigeria: the role of the National centre for Agricultural Mechanisation. *Journal of Agricultural Engineering and Technology (JAET)*, 17(2).

F.A.O. 2008. Agricultural mechanisation in Africa: time for action. Food and Agricultural Organisation, *www.unido.org/fileadmin/.../agricultural_mechanisation_in_Africa.pdf*, [Retrieved Sept. 13, 2013].

NCAM (2013) National Centre for Agricultural Mechanisation (NCAM) web site.

Odigboh E. U. 2000. Mechanisation of the Nigeria agricultural industry, pertinent notes, pressing issues, pragmatic options. A public lecture delivered at the Nigeria Academy of Science, International Conference Centre, Abuja. April 15.

Ruttan V. W. 2013. Growth stage theories and agricultural development policies. *ideas.repec.org/a/ags/ajaeau/22652.html*, [Retrieved September 12, 2013].

Todaro, M. P. and S. C. Smith. 2009. Economic development. Pearson Addison Wesley. 8th edition. http://wps.aw.com/aw_todarosmit_econdevelp_8/4/1111/284635.cw/index.html, [Retrieved September 9, 2013].

CHAPTER 3:

Rural innovations and knowledge systems development and dissemination among cassava cooperative farmers in Southern Nigeria

Edwards Adeseye Alademerin

1.
INTRODUCTION AND JUSTIFICATIONS FOR FARMER'S INNOVATION AND INDIGENOUS KNOWLEDGE EXCHANGE SYSTEM

Innovation as a basis for economic development has been emphasised by Economic Commission for Africa and the United Nations Millennium Project (2005). Innovation can be defined as all the scientific, technological, organizational, financial, and commercial activities necessary to create, implement, and market new or improved products or processes (OECD, 1997).

An innovation is an idea, behaviour, or object that is perceived as new by its audience and usually, it is seen as an end result of the urge to improve livelihoods and make them more sustainable for economic uses by man. "Innovations around the most basic humanitarian needs are always top priority, but helping communities get back on their feet and start to function as economically-independent units is just as important as emergency aid support" (Michael Pritchard MBE, 2013)

Innovation as a concept is used here in this paper with indigenous knowledge (IK) system in relation to agricultural productivity particularly by cassava farmers in the rural areas of southern Nigeria. Cassava farmers have been able to improve their production level and income through innovations and the development of indigenous knowledge exchanges among fellow farmers within their neighbourhood and beyond. This is borne out of the fact that interaction among fellow farmers is itself a poverty reduction strategy

The importance of this traditional knowledge for the protection of biodiversity and the achievement of sustainable development is slowly being recognized internationally (Gadgil *et al*, 1993). The poor farmers themselves appreciate their situations better and must therefore be involved in the design of both formal and informal means to better their lot in their immediate communities. Such improvements can be in the types of crop varieties grown, diseases and pests control, processing and storage techniques, sales etc. It is through these and several other approaches that they can be involved in the true concept of sustainable livelihood. *"A livelihood comprises the capabilities, assets (including both material and social resources), and activities required for a means of living. A livelihood is sustainable when it can cope with and recover from stresses and shocks and maintain or enhance its capabilities and assets both now and in the future, while not undermining the natural resource base"* (DFID, 1997).

Cultures from all over the world have developed different views of nature throughout human history. Many of them are rooted in traditional systems of beliefs, which indigenous people use to understand and interpret their bio-physical environment (Iaccarino, 2003). The ultimate aim of understanding our bio-physical environments and nature is to be able to tackle poverty once and for all. DFID's sustainable livelihoods (SL) approach aims to increase the agency's effectiveness in poverty reduction in two main ways: the first is by mainstreaming a set of core principles which determine that poverty-focused development activity should be people-centred, responsive and participatory, multi-level, conducted in partnership, sustainable, and dynamic. The second is by applying a holistic perspective in the programming of support activities, to ensure that these correspond to issues or areas of direct relevance for improving poor people's livelihoods (Krantz, 2001).

The nature and intensity of innovations have over the years improved farmers' level of awareness, sustainable livelihoods and consequently reduced the poverty level amongst their households. Being aware of improved production and storage practices will in turn improve their income, health and standard of living. There is no doubt that effective participation by the poor in grassroots projects initiated by government and other private agencies all over the world enhances overall project performance. Indigenous knowledge is often well displayed in such instances and a clear example here is the cassava multiplication programme (CMP) of International Fund for Agricultural Development (IFAD) initiated in Nigeria in 1985. The potential role of indigenous knowledge (IK) in improving agricultural performance is widely recognised in developing countries (Hart 2007). Transfer of IK from generation to generation is mostly done through oral tradition or by demonstration.

Knowledge management (KM) has been successfully adopted by many organisations in order to build their competitive strength and achieve a sustainable growth pattern (Ichijo Nonaka 2007). KM practices in closed systems or formal organisations are likely to be more successful than those in informal systems or open systems because they have formal structures and rules to which members of organisations adhere (Mosia and Ngulube, 2005).

Knowledge assets are key elements that facilitate knowledge creation processes. Those assets include:

- experiential (i.e. skills and know-how)
- conceptual (i.e. concepts, designs and methods)
- systemic (i.e. technological platforms, manuals and patents and licences)
- routine (i.e. know-how in daily operations) (Lwoga, Ngulubeand Stilwell, 2010).

The beauty of knowledge is in the development and preservation for future uses. However, KM should also be applied in the rural communities of developing countries for equitable and sustainable development because knowledge is an important resource for socio-economic growth. Rural communities have an extensive base of IK which is at risk of becoming extinct if appropriate measures are not taken to manage it. KM can be used to manage and share IK in communities that desire to achieve development agendas (Ngulube, 2008).

Many indigenous populations have relied for centuries or even millennia on their direct environment for subsistence and autonomy. Over time, they have developed a way in which to manage and use their resources that ensures their conservation into the future....... Natural resource management is based on shared meanings and knowledge (Berkes, 1993). Innovations are essential in traditional societies because of the various interests in improving as well as preserving their own social, cultural and environmental stability. In general, traditional knowledge systems adopt a more holistic approach, and do not separate observations into different disciplines as does Western science (Iaccarino, 2003).

2.
THE THEORIES AND MODELS OF INNOVATION AND INDIGENOUS KNOWLEDGE SYSTEMS

As it has been indicated earlier, the basis of innovation is for the improvements of the environments economically for the benefits of mankind. Usually, the long term effect is to make livelihoods more sustainable for mankind. Innovation as a concept can as well be referred to as a process of an enduring change since it normally involves a group of people with a mission that are interconnected and constantly interacting among one and another and serving as agents of positive change.

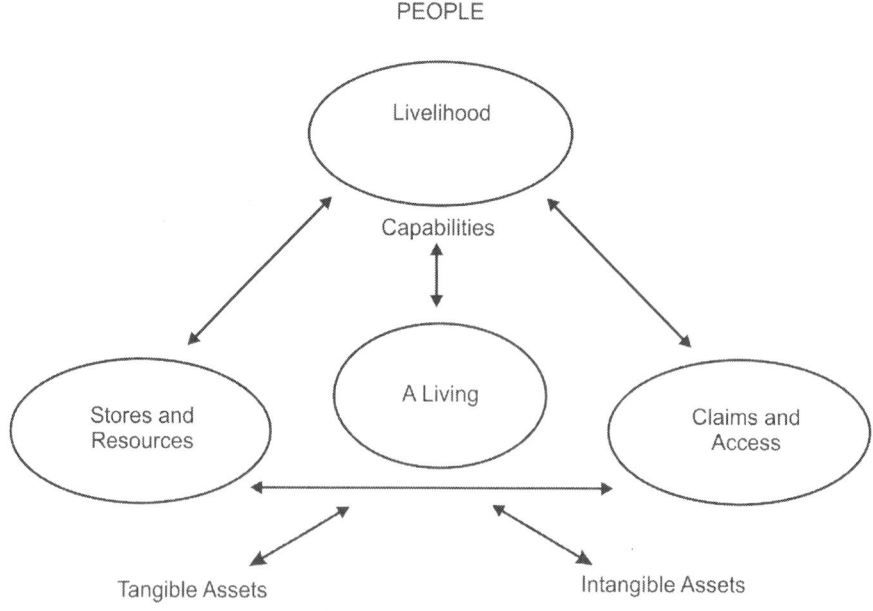

Source: Krantz (2001).

Fig: 1. UNDP'S approach to promoting sustainable livelihoods

The sustainability of livelihoods becomes a function of how men and women use asset portfolios on both a short- and long-term basis. Sustainable livelihoods are those that are:

- able to cope with and recover from shocks and stresses through adaptive and coping strategies;
- economically effective;
- ecologically sound, ensuring that livelihood activities do not irreversibly degrade natural resources within a given ecosystem; and
- socially equitable, which suggests that promotion of livelihood opportunities for one group should not foreclose options for other groups, either now or in the future.

Indigenous ways of knowing are based on locally, ecologically, and seasonally contextualized truths. In contrast to the aspirations of some Western scientific traditions for universal truths, Indigenous epistemologies are narratively anchored in natural communities. Those natural communities are characterised by complex kinship systems of relationships among people, animals, the earth, the cosmos, etc. from which knowing originates (Ermine, 1995: 101-112).

According to Robbinson (2009), "Diffusion of Innovations offers three valuable insights into the process of social change:

- What qualities make an innovation spread?
- The importance of peer-peer conversations and peer networks.
- Understanding the needs of different user segments.

Based on these distinctions, one can also distinguish between two types of innovations: cumulative innovation motivated by the need for improvements that has been identified through routinized activities, and discrete, independent development that often indicates the beginning of a new technological paradigm (Dosi and Nelson, 1994; Klevorick et al., 1995).

Pressures from a societal problem can bring about an innovation and gradual positive response from the people particularly when it involves a local means of communication. Production of folk drama and audio-visual such as in the Kenya Wood fuel development programme in Kakamega district was also a successful exercise. Local actors and comedians were employed to stage an amateur drama incorporating local saying and songs. The experiences gained in collaborative experimentation with respect to tree planting and management with a limited number of farmer groups were fed into a drama which is staged on market days. The drama provided a reflection on the woodfuel problem, tells about the experimentation done by the initial groups, and encourages participants to develop their own ideas as to how the situation can be improved (Kenyan Wildlife Agricultural Programme (KWAP), 1991).

Hammersmith (2007) quoting from previous works agreed that Indigenous communities generally describe Indigenous knowledge as:

- practical common sense based on the teachings and experiences passed on from generation to generation.
- knowing its home country. Indigenous knowledge covers knowledge of the environment - snow, ice, weather, resources - and the relationships among things.
- holistic; it cannot be compartmentalised and cannot be separated from the people. It is rooted in the spiritual health, culture and language of the people. It is a way of life.
- a traditional authority system; setting out the rules governing the use of resources - respect, an obligation to share. It is dynamic, cumulative and stable. It is truth.
- a way of life - wisdom is using traditional knowledge in 'good' ways. It means using the heart and the head together. It survives because it comes from the spirit.
- giving credibility to people.
- serving community needs and interests first.
- having the potential to realise that the real contributions of local and traditional knowledge incorporate knowledge of the ecosystem.
- relationships and a code of ethics, govern the appropriate use of the environment.
- recognising that this code of ethics includes rules and conventions promoting desirable ecosystem relations, human-animal interactions and even social relationships.
- enabling traditional knowledge to articulate with non-traditional knowledge to form a rich and distinctive understanding of life and the world.

Quoting from Mayor (1994), Hammersmith (2007) indicated that "the world's Indigenous people possess immense knowledge of their environments, based on centuries of living close to nature. He points out in an Opening Address (Mayor 1994: 1-6) to a 1994 UNESCO Lifelong Learning Conference in Rome, that living in and from complex ecosystems, these people have an understanding of the properties of plants and animals, the functioning of ecosystems and the techniques for using and managing them that is particular and often detailed. His address continues that in rural communities in developing countries, locally occurring species are relied on for, sometimes all, foods, medicines, fuel, building materials and other products.

In addition, he says that peoples' knowledge and perceptions of the environment, and their relationships with it, are often important elements of their cultural identity. Most Indigenous people make use of traditional songs, stories, legends, dreams, methods and practises as a means of transmitting specific human elements of traditional knowledge. Sometimes they are preserved in artefacts handed from one generation to the next. In the *context* of Indigenous knowledge systems, there is usually no real separation between secular and sacred knowledge and practise. They are one and the same. In virtually all of these systems, knowledge is transmitted directly from individual to individual.

Onu (1990), opined that participation in various farmer social organization is generally considered an important variable that enhances farmers' adoption of new practices Nigerian farmers who participate actively in the life of their communities, membership in, and leadership of social organizations such as former co-operative societies, thrift and credit societies, church organizations, social clubs, age grades, village council and contact farmership etc. are more likely to be exposed to communication message that are related to farm innovations and adoption more than their other counterparts due to group dynamic effects.

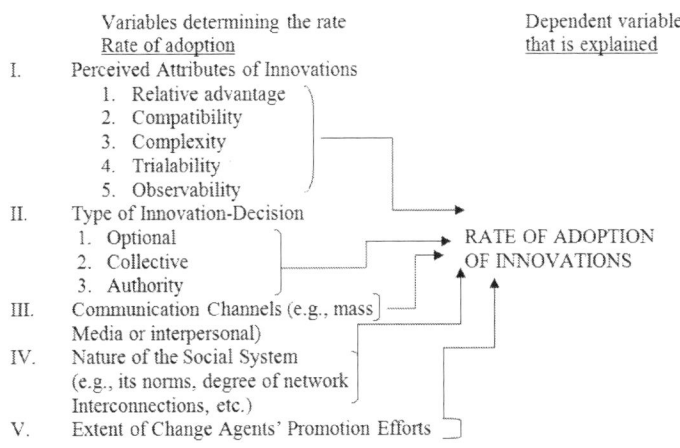

Source: Rogers (1995).

Fig. 2. Variables determining the rate of adoption of innovations

3.
RESEARCH AND DEVELOPMENT REGARDING INFORMATION AND KNOWLEDGE DISSEMINATION AMONG CASSAVA CO-OPERATIVE FARMERS

Researches and development into cassava crop are not recent in Nigeria and in Africa as a whole. Government's intervention and efforts of non-government organizations (NGOs) in the cassava sub-sector have led to a number of measures and interventions both nationally and continentally since 1970's by International Fund for Agricultural Development (IFAD), World Bank, Food and Agriculture Organisation (FAO) etc. IFAD has a mandate, unique among such institutions to combat hunger and rural poverty, focusing in particular on the low income food deficit countries that receive well over 80% of its loans. It directs funding and mobilizes additional resources on concessional terms to finance rural development projects involving the world's poorest populations; small farmers, landless poor, artisan, fishermen, nomadic herdsmen and poor rural women. The aim is to increase their food production, raise their incomes, improve their health, nutrition and educational standards, and ensure their well-being in a sustainable and environmentally safe way (IFAD, 1998).

Cassava is a major source of dietary energy for low income consumers in many parts of tropical Africa, including major urban areas. In line with its adaptability and tolerance, the Food and Agriculture Organisation (FAO, 1998) indicated that Cassava (Manihot esculenta) is grown over a harvested area of 16 million hectares with a global production of 161 million tonnes (1995), cassava is one of the most efficient crops in biomass production. In comparison with many other crops, it excels under sub-optimal conditions and can withstand drought conditions. The five highest producers are Brazil (16%), Nigeria (13%), the Democratic Republic of Congo (11%), Thailand (10%) and Indonesia (10%) of global production.

In a broader sense, it targets the "poorest of the poor" and it ensures food security for all. The Cassava Multiplication Programme (CMP) of the International Fund for Agricultural Development (IFAD) is one of the numerous agricultural development programmes aimed at increasing production and is therefore essentially growth – oriented, with the understanding that such growth will eventually induce development on the part of farmers, reduce hunger among the larger populace and increase their economic status. Romanof and Lynam (1992) argue also that cassava plays a famine prevention role; where cassava is widely grown, famine rarely occurs because cassava provides a stable base to the food production system. This is one of the reasons why IFAD focuses on project – type interventions to alleviate human sufferings in food availability through a collective efforts and participatory approach.

According to FMANR 1997,v) "the IFAD-assisted Cassava Multiplication Programme (CMP) was conceived following severe attack on cassava crop by two alien pests, cassava mealy bug (CM) and cassava green spider mite (CGM), in the late 1970s and early 1980s and the resultant decline in production. At the instance of the Federal Government of Nigeria (FGN), the IFAD approved a loan of United States dollars $12.05 million in 1986 for cassava improvement programme in the Southern State of Nigeria.

According to IFAD 1994:4), "Cassava is a crop of the poor and occupies mainly agriculturally marginal environments. These and other features endow it with a special capacity to contribute to food security, equity, poverty, alleviation and environmental protection". Improving this crop is a way to direct various benefits toward the poorest of rural populations. Several efforts have been directed towards this by the FGN in contributing to efforts to distribute improved cassava planting materials since the inception of the CMP of IFAD in 1986. Since the development and gradual introduction of the tropical manihot selections (TMS) i.e. new cassava varieties, hectare yield has improved significantly over the use of the old varieties. According to FMANR (1999:3), 'the new varieties have demonstrable high yield potentials of 30 – 35 metric tonnes per hectare well above the indigenous varieties which yield less than 10 tonnes per hectare. The major problem with the breeding programme was the slow pace of development and long delay in the release of new varieties'.

The new TMS varieties are gradually spreading to all the nooks and crannies of the rural areas and the relevance of farmers on fellow farmers is high regarding the spread and acceptance by them. Since the TMS is a new innovation and idea, the readiness to accept them and put them into practice varies from farmers to farmer depending on each farmer's previous experience with new ideas; the personality of the farmer and the amount of land and other resources available at his disposal.

Frequently, when ideas about the TMS are shared with friends, seed materials are exchanged; new products gain recognition usually along trading routes, local markets and farmers' forum meetings. Agricultural ideas relating to the TMS have also enjoyed much dissemination from the extension workers attached to the ADPs in growing areas. According to FMANR (1997:5), "the extension activities of the ADPs led to the rapid adoption of improved cassava varieties and expansion of the area planted to cassava in the major cassava producing states with the result that the new varieties are progressively replacing the traditional ones. In most of the developing countries, the reliance of farmers on fellow farmers is high regarding research and extension services than on government extension officers. At local and regional level, the ability of farmers' organisations to influence research and extension depends on the structure of the system itself.

In a decentralized system, farmers' organizations tend to have better access to individuals, within the system as they are physically closer at hand. This is very common at village and rural levels in the cassava growing states of Nigeria. Friends and relations constantly bring about innovations and exchange ideas, seeds of crops, planting materials of cassava among themselves and this system seems to be much more efficient than the conventional government extension services.

Spontaneous diffusion of technologies (that have been proved successful) occurs frequently when ideas are shared with friends, seeds and materials are exchanged, and new products gain recognition along trading routes. Local markets or meetings may be important venues for sharing agricultural ideas. A great variety of method -drama, song, Jokes – may be important locally to carry agricultural messages. There is now increasing recognition of the importance of indigenous common networks (Box, 1989; Simpson, 1994).

In line with the above, Veldhuizen et al (1997) agree that farmers may indeed play an important role and take over responsibility in spreading experience on agricultural innovations. This is known as farmer-based extension. Farmers distinguish land races by local names which are often descriptive of the physical characteristics of the plant, such as the colour of certain parts, height and canopy size, yield potential, bulking period etc. Such names may also describe the original source of the genotype, such as the place or institution from which it was introduced for the first time, or the individual who brought it originally. The local names may also be an indication of an event which coincided with the introduction of the genotype in the village (Jones, 1959). Examples are Maadan, Olengasiya, Oreke, Mafamipa, Oko Iyawo, Dalejoro etc. in Western Nigeria. Involving farmers' organization in the innovation, adoption and spread of some varieties of crops and technologies has proved very successful in some other parts of the world. Tables 1 and 2 below represent the findings from a research work conducted on knowledge about the new cassava varieties in Southern Nigeria between 1997 and 2001:

Table 1. Distribution of Respondents by Traditional and Social Status

S/N	Traditional/Social Status	Frequency (No.)	Percentage (%)
1.	Village head	11	3.3
2.	A sectional head	41	12.4
3.	A chief	14	4.2
4.	An ordinary member	195	58.9
5.	A politician	38	11.5
6.	A club member	19	5.7
7.	A religious leader	13	3.9
8.	A musician	0	0
	TOTAL	**331**	**100.0**

Field survey, (2000).

Table 1 show that 11 (3.3%) of the cooperative farmers were village heads, 41 (12.4%) were sectional heads, 14 (4.2%) were chiefs and 195 (55.9%) were ordinary members. The table also shows that 38 (11.5%) were party politicians, 19 (5.7%) were members of social clubs while 13 (3.9%) were religious leaders. None of the farmers was a musician.

Table 2: Distribution of Respondents Regarding Knowledge About TMS

S/N	Source of Knowledge	Frequency (No.)	Percentage (%)
1.	Friends	11	3.3
2.	Agricultural bulletins	41	12.4
3.	Newspapers and Magazines	14	4.2
4.	Extension workers of ADP	195	58.9
5.	From Radio	38	11.5
6.	Market/Trading routes	19	5.7
7.	Cooperative Societies	13	3.9
8.	From government	0	0
9.	NGOs	4	1.2
10.	OFAR	25	7.6
	TOTAL	**331**	**100.0**

Field survey, (2000).

Table 4 shows that 38 (11.5%) of the cooperative farmers got their knowledge about TMS from friends, 1 (0.3%) from agricultural bulletin, while 1 (0.3%) from newspapers and magazines. The table also shows that 256 (77.3%) got their knowledge about TMS from extension workers of ADP, 6 (1.8%) from cooperative societies, 4 (1.2%) from non-governmental organizations and 25 (7.6%) through On-Farm Adaptive Research (OFAR). None got the knowledge from radio programmers, market and trading routes and also from government sources

4
STRENGTHS AND WEAKNESSES OF THE LINKAGES AND NETWORKS AMONGST COMMUNITIES TO PROMOTE INCLUSIVE INNOVATIONS FOR SUSTAINABLE DEVELOPMENT

Cooperation and active participation by farming communities in government programmes are variables required for rapid development and transformation of such communities particularly in tropical Africa. Linkages and networks amongst communities are taken to mean cooperative movements among the rural farming communities in this paper. There is no doubt that functional cooperative movements are synonymous with rapid agricultural transformation of the rural areas.

According to Bello (1996:41) ``Co-op is a natural phenomenon engendered and nurtured by inbuilt survival instincts. This has been demonstrated by all living creatures – man, animal and plants – in one aggregation or the other...... On the spiritual plane, religious concepts were formulated, believed and acted upon by people of such sect only because there is a mutual agreement to share and worship together, i.e. co-operate`` In getting the desired cooperation, the traditional institutions as well as local societies and unions must be adequately empowered to maximally function. Traditional and customary institutions in the rural areas are given due considerations because of their significant roles they play too in adoption and spread of any innovation. By customary institutions, we refer to those relationships that have long been the basis of social organization. These include kinship networks, tenure rules, local concepts of "the community" the rules governing gender, relationships, local criteria determining who has authority and how decisions get made etc. these are the rules institutions that are most deeply band into the organization of rural life, and which make most sense to, and have most hold over rural people (Moorhead and Lane, 1993).

The history of co-operative (co-op) movement is as old as man. The present efforts of scientists and technologist worldwide towards globalization are as a result of everyday co-operation which is fast yielding good result. The all-encompassing benefits of co-operatives are no doubt evident in our everyday life and the continual exploitation of these benefits will make life more meaningful to all. The impact of all these is the newer concept of ``Global neighbourhood`` instead of the old term of ``Global village`` which has made the world at the reach of all and sundry with the press of a button from an electronic terminal of a computer set.

In rural areas, informal co-operation is demonstrated when resources are pooled by individuals to clear farm land, cultivate, harvest, process farm produce, hunt animals, build houses or engage in social and cultural activities. One principle that stands out clearly in co-operation (be it economic, non-economic, social or cultural) is that it results in the attainment of a desired objective of the relevant units (Nweze, 1997:1).

In Bassar, Togo (two years after the facilitator of World Neighbours left), 12 village communities were continuing to meet annually to analyse and evaluate the previous season's experiments and schedule the research agenda for the coming season. At this meeting they also choose a limited number of delegates to make the rounds of various agricultural development programmes and research stations, actively seeking new ideas and technologies and to reports back to their communities (Gubbels, 1988).

Mosse (1993), for instance, argues on the basis of evidence from the Kribhco rain-fed farming project in Western India, that village appraisal and planning initiatives that did not build on existing authority structures were likely to be obstructed by village leaders. Quoting the works of Rivera-Cucicanqui (1990), Bebbington et al (1994) wrote on a similar lesson which comes from the experience of an NGO research and extension project working in the Bolivian Andes, which tried to create local organizations. These organizations were intended to be the village – level counterpart for the programme. Consequently, the organizations created by the NGO project attracted the interest of these young adults, who saw them as a means of gaining authority that traditional rules did not allow, and the project unwittingly created parallel authority structures in communities that essentially pitted the young adults against the old. In this case, the attempt to create and then work through local organizations led to conflict in communities rather than a more farmer responsive and effective research/extension programme.

Balbariho (1994) reported in Veldhuizen et al (1997) stated that farmer experimenters join annual research review meetings at research institutes, as done by FARMI in the Philippines. Ability of these farmers to participate in formal meetings and contribute in adoption and spread of innovations are as a result of their knowledge of culture, local and traditional documentation.

Documentation by farmers of their experiences in FPR activities plays a very important role in spreading the ideas; it helps to make these experiences accessible for other villages and areas. Traditionally, experiences are kept in the memories of the people, or are translated into songs, jokes or simple drama for the benefit of future generations; this is a local communication means of preserving culture and farming.

The dominant roles that cassava farmers play in tropical Africa in areas of staple food supplies cannot be overemphasised. The rural nature of the cassava farmers makes poverty to be prevalent among them. There is growing recognition that poverty is not only about income, but about social risks such as discrimination, unequal distribution of resources and power in households and limited citizenship (CPRC, 2008). "Cassava is a crop of the poor, and occupies mainly agriculturally marginal environments. These and other features endow it with special capacities to contribute to food security, equity, poverty alleviation and environmental protection" (CIAT, 1999:4).

In fact, cassava is also an important source of cash income for poor farmers, as well as prosperous ones. Apart from its potential as a source of increasing total income from agriculture, cassava may also play a role in achieving a more egalitarian pattern of income distribution and stability thus;-

Cassava may be harvested and sold in small amounts at frequent intervals. It can provide a steady flow of cash income over many months

It is relatively easy to combine with domestic chores and other income-earning activities on a daily or weekly basis. This is an advantage, especially for rural women, who may find it difficult to specialize, even temporarily, in harvesting, processing, and or marketing a single crop, and who lack the working capital or social position to mobilize the labour of others (Guyer, 1980).

Cassava may be harvested and sold in bulk to take advantage of favourable prices or provide producers with liquidity to finance lumpy consumption of investment outlays. The in-ground, self-storing capacity of cassava permits flexibility in harvesting and marketing which can be advantageous to specialized, fully commercialized producers, as well as to smaller, diversified, and/or partially commercialized ones.

Berry (1993) reported that in Nigeria and Congo Democratic Republic, and elsewhere, there are both large and small-scale farms on which cassava is grown entirely for sale, by both full – and part-time farmers. In Nigeria for example, in the mid-80s, rising food crop prices and the oil recession reduced economic opportunities outside of agriculture, and many people (including wage and salary earners, professionals, traders etc.) established cassava farms. Such investors used working capital from other sources to hire labour for land preparation, planting and initial weeding; then sold the crop in the ground to buyers who assumed full responsibility for further weeding, harvest and sale.

These are indications that the crop is indeed a poverty alleviation crop for producing households.

5
RECOMMENDATIONS ON WAYS TO HARNESS INDIGENOUS KNOWLEDGE SYSTEMS TO PROMOTE INCLUSIVE INNOVATIONS FOR SUSTAINABLE DEVELOPMENT

Innovations at all levels of human endeavours are mostly directed at making life pleasurable and sustainable for mankind in day to day interactions within their immediate and outer environments. Innovation should benefit all and sundry within a society in diverse ways to assist in combating the menace of poverty. The followings are hereby recommended:

Awareness and mobilisation efforts of National governments should be intensified in the areas of the introduction of new varieties of crops particularly staples and also helping to make them available to farmers at subsidised rates

Development of IKM system among rural communities through a cost effective strategy and information retrieval system.

Government agencies and NGO's should assist in encouraging diversified livelihoods of the rural people by helping them to understand their bio-physical environments through waste to wealth strategy and entrepreneurship education.

Traditional institutions should create awareness and encourage people to belong to social and religious organisations as these have been found to increase innovation techniques and adoption rates.

Local agencies like the cooperatives and farmers unions should ensure that new innovations, technological and scientific feats build on existing indigenous knowledge.

Establishment of farm settlement schemes and encouragement of improved processing techniques of cassava into various products e.g. - cassava flour to gradually replace wheat in baking of bread etc.

Wide uses of various mass media – TV, radio, posters, hand bills, bill boards etc. in the dissemination of various research results among farming communities should be encouraged.

REFERENCES

Alademerin, E. A. (2001). "Evaluation of the impact of cassava Multiplication Programme of the International Fund for Agricultural Development on Cooperative Farmers in Southern Nigeria". A Ph.D. Thesis submitted to the Department of Vocational Teacher Education, University of Nigeria, Nsukka.

Bebbington, A. 1999. "Capitals and Capabilities: A Framework for Analysing Peasant Viability, Rural Livelihoods and Poverty", *World Development,* 27(12), 2021-2044.

Bello, S. B. (1996). Essentials of Co-operative Administration for Primary Societies in Nigeria. Lagos: Lagos State Printing Corporation, Ikeja.

Berdeque, J. A. (1990). "NGOs and farmers: Organisations in Research and Extension and Extension in Chile". London, UK: ODI/AGREN Network Paper 19.

Berkes, F. (1993). Traditional ecological knowledge in perspective. In Inglis JT (ed.) Traditional Ecological Knowledge: Concept and Cases, pp 1–9. Ottawa, Canada: International Program on Traditional Ecological Knowledge and International Development Research Centre.

Berry, S. S. (1993). Socio-Economic Aspects of Cassava Cultivation and Use in Africa: Implications for the Development of Appropriate Technology. COSCA Working Paper No. 8 IITA, Ibadan. Collaborative Study for Cassava in Africa.

Box, L. (1989). "Virgillio's Theorem: A Method for Adequate Research" In Chambers, R; Pacey, A. and Thrupp, L A. (eds.) Farmer First: Farmer Innovation and Agricultural Research, London Intermediate Technology Publications.

Chambers, R. (1993). Challenging the Professions. Frontiers for Rural Development. London: IT Publications.

Chronic Poverty Research Centre (CPRC, 2008). The chronic poverty report 2008-9; Escaping poverty traps. Chronic Poverty Research Centre. Manchester, UK.

DFID. *Sustainable Livelihoods and Poverty Elimination: Background Briefing.* (www.ids.ac.uk/livelihoods.html). [Accessed November 10, 1999]

Emery, Alan R. And Associates (1997). "Guidelines For Environmental Assessments and Traditional Knowledge," *Knowledge Of The World Council Of Indigenous People, (Daft) A Report From The Centre For Traditional Knowledge.* Ottawa: WCIP.

Ermine, Willie, (1995). "Aboriginal Epistemology" in Batiste, Marie and Jean Barman (eds.) *First Nations Education In Canada: The Circle Unfolds*, Vancouver, UBC Press.

Federal Ministry of Agriculture and National Resources (FMANR), (1998). Development of Cassava in Nigeria. FMANR.

Food and Agriculture Organisation (FAO) (1985). Guide to extension Training, Rome: FAO Publication Unit.

Food and Agriculture Organisation (1998). formulation reports: Roots and Tubers Expansion programme. Rome. FAO.

Gadgil M, Berkes F, Folke C (1993). Indigenous knowledge for biodiversity conservation. *Ambio*, 22(2-3), 151–156.

Guyer, J. (1980). Food, Cocoa and the division of labour by sex in five West African societies. *Comparative Studies in Society and History*. 22(3).

Gubbels, P. (1988) "Peasant farmer agricultural self-development: the World Neighbours experience in West Africa in : ILEIA Newsletter; 4 (3).

Hammersmith, J.A (2007) Converging Indigenous and Western Knowledge Systems: Implications for Tertiary Education, University of South Africa. Unpublished D. Ed. thesis submitted.

Ichijo, K. Nonaka, I., (2007). 'Knowledge as competitive advantage in the age of increasing globalization', in K. Ichijo I. Nonaka (eds.), *Knowledge creation and management: New challenges for* managers, pp. ix-xiv, Oxford University Press, Oxford.

IFAD (1998). IFAD Annual Report 1998. Rome. IFAD Publication Unit.

IFAD (1998). A Call to action. Building on IFAD's Experience: Lessons from Sub-Saharan Africa.

IFAD consultation in West and central Africa Region Consultation on Global Cassava Development Strategy. Accra-Ghana1-3 June 1999.

IFAD (1998). IFAD Annual Report 1998. Rome. IFAD Publication Unit.

IFAD (1998). A Cell to Action. Building on IFAD's Experience: Lessons from Sub-Saharan Africa.

Kawagley, O. (1995). *Yupiaq Worldview*. Prospect Heights, IL: Waveland Press.

Krantz, L (2001). The Sustainable Livelihood Approach to Poverty Reduction: An Introduction. Swedish International Development Cooperation Agency. Division for Policy and Socio-Economic Analysis.

KWAP (1991). Participatory agro-forestry extension; a field handbook, Kenyan Wood fuel and Agroforestry programme, Kenya.

Iaccarino M (2003). Science and culture. EMBO Rep 4: 220–223.

Les Robbinson (2009). A summary of diffusion of innovation. [Avalable at]: http://www.enablingchange.com.au/Summary_Diffusion_Theory.pdf

Mayor, Federico (1994) in Opening Address to UNESCO Conference on Lifelong Learning, Rome, Nov. 30, 1994.

Mosia, L.N. Ngulube, P., (2005). 'Managing the collective intelligence of local communities for the sustainable utilisation of estuaries in the Eastern Cape, South Africa', *South African Journal of Libraries and Information Science*, 71(2), 175-186.

Mosse, D. (1994). 'Authority, Gender and Knowledge: Theoretical reflections on the practice of participatory rural appraisal', *Development and Change*. 25(3), 497- 526.

Mosse, David. (1995). 'Local Institutions and Power: The history and practice of community management of tank irrigation systems in South India', in Nici Nelson and Susan Wright (eds.) *Power and Participatory Development: Theory and Practice.* IT Publications, London.

Nelson, R. R. (1993). *National Innovation Systems - A comparative analysis*. New York and Oxford: Oxford University Press.

Ngulube, P., (2003). 'Using the SECI knowledge management model and other tools to communicate and manage tacit indigenous knowledge', *Innovation*, 27(1), 21-30.

Nweke, F. I. (1998). Production prospects for cassava in Tanzania. COSCA working paper No. 16 Ibadan. COSCA, IITA.

Nweze, N. J. (1997). Essentials of cooperative Economics. Lagos; A. Johnson Publishers.

Okali, C. and Berry, S. (1985). Alley farming in West Africa in comparative perspective. Working paper No 11 Boston, African American issues Centre.

OECD. (1997). The Measurement of Scientific and Technological Activities, Proposed Guidelines for Collecting and Interpreting Technological Innovation Data. Paris: Organization for Economic Cooperation and Development.

Onu, D. O. (1990). Factors associated with small scale farmers adoption of improved soil conservation techniques under intensified agriculture in Imo state, Nigeria. Unpublished PhD thesis, UNN.

Romanoff, S. and Lynam, J. (1992). Commentary: Cassava and Africa food security: some ethnographic examples. Ecology of food and Nutrition 27.

Rogers, E. M. (1995). Diffusion of Innovations. Fourth edition, The Free Press New York . [Available at]: http://www.d.umn.edu/~lrochfor/ireland/dif-of-in-ch06.pdf

Rogers, E.M. (2003). Diffusion of Innovations (5th Edition). New York: The Free Press.

Rivera – Cucicanqui, S. (1990). Liberal Democracy and Ayllu Democracy in Bolivia: the case of Northern Potosi'. Pp 97 – 121 in J. Fox (ed.) 1990a. Rural Democratization. London. Frank Cass.

Simpson, 1994). "The lifeblood of Agricultural change". In ILETA Newsletter: 10 (1): 16.

UN Millennium Project. (2005). *Innovation: Applying Knowledge in Development*. London: Task Force on Science, Technology and Innovation, Earthscan.

United Nations Educational, Scientific and Cultural Organization, (2002). Universal Declaration on Cultural Diversity. Paris, France: UNESCO.

UNDP. *Sustainable Livelihoods Approaches in Operations: A Gender Perspective.* [Available at]: (www.undp.org/sl.htm)

UNDP. *Policy Analysis and Formulation for Sustainable Livelihoods.* [Available at]: (www.undp.org/sl.htm)

USDA (1981). Food problems and prospects in sub-Saharan Africa. Washington, D. C.

Veldhuizen, L. V.; Waters-Bayer, A. and De Zeeuw, H. (1997) Developing Technology with Farmers: A Trainer's Guide for Participatory Learning. London: Zed Books Ltd.

World Bank, (1987) Accelerated development in sub-Saharan Africa Washington, D. C.

World Bank, (1984). Towards sustained development for sub-Saharan Africa Washington, D.C.

CHAPTER 4:

The role of Innovation in Development – could lessons be drawn for Ethiopia?

Bedru B. Balana

1.
INTRODUCTION

In the first chapter of his book entitled 'Economic Development', Michael Todaro, a leading contemporary Development Economist, asked the seemingly known but a question that remains elusive for many – *"What do we mean by Development?"* (Todaro, 1998, p.13). Development has traditionally (esp. in the 1950s and 1960s) been defined and measured by an increase in the GNP per capita. However, in the last 2 decades or so (since the 1980s) development has been conceived as a multidimensional process involving major changes in the social, economic, attitudinal, and institutional changes that encompass the social system as a whole instead of a simple per capita income growth. Todaro (1998) summarised the three objectives of development as to: increase the availability and widen the distribution of basic goods (food, shelter, health and protection); raise levels of living (e.g. employment opportunities and better education); and increase the range of economic and social choices.

As the term 'development' may mean different things to different people, the importance of having a common perspective and an agreed measurement criteria should be emphasised, without which we would not be able to determine policy measures to tackle the problem. Now a consensus is emerging that most economists, other scholars, international organizations and national governments recognize that development is multidimensional (at least development and growth are not the same). Just as there has been lacks in common understanding of the term 'development', there appears no consensus on the remedies for the problems of development. Economist, particularly, in the second half of the 20th century have postulated myriads of remedies for the problems of development. These vary from the Hard-Domar model of 'financing gap' to the Nurkse's 'balanced growth' (a model of the synchronized application capital to a wide range of industries) and Hirschman's 'unbalanced growth' (a theory that focuses on key industries to maximize forward and backward linkages) to those considering foreign aid as panacea for development.

With regard to the emphasis on the specific factors for development, economists' view vary significantly – some advocate existence of natural resources such as oil reserves and fertile land as the major factors for the development. For some, lack of physical capital and infrastructure is considered as the key bottleneck for development. Few others put their blame on geographical and weather factors as the impediments for development. In recent times there has been an increasing tendency to recognize that education, human capital formation, technology and knowledge as the key element of development. Our interest in this paper is to examine the role that knowledge, technology and particularly innovation plays in countries' development process.

2.
INNOVATION FOR DEVELOPMENT

Innovation – understanding the concept

The concept of 'innovation' has been defined in slightly various ways. The OECD, for instance, defined innovation as "the implementation of a new or significantly improved product (good or service), or process, a new marketing method, or a new organizational method in business practices, workplace organization or external relations" (OECD and European Communities, 2005, p. 46). Onodera and Kim (2008, p. 112) think that innovation "is about the successful exploitation of new ideas and the invention, development and commercialization of new technologies, services, business models and operational methods. Innovation is thus related to a process connecting knowledge and technology with the exploitation of market opportunities for new or improved products, services and business processes compared to those already available on the market."

The UNESCO (2009) – Summary report on 'Innovation for Development – Converting Knowledge to value' seems inclined to associate innovation with 'information in science and technology. According to this report "Innovation relates to the introduction of a new idea, product or process to a user or user group, and refers particularly to the transfer and application of knowledge, R&D and information in science and engineering, often embodied or embedded in technology. Innovation is a social and economic as well as a technical process, and knowledge or technology transferred does not need to be absolutely new, nor does it consist only of 'formal' knowledge. Innovation more commonly involves *incremental* rather than radical change, based on engineering research, design, development and 'learning by doing', rather than scientific research."

In a more broad sense, innovation can be conceptualized as the ***creative use of knowledge*** to allow individuals, groups, organizations and governments – "to go farther, faster, deeper and cheaper" (Friedman, 1999). In most instances, innovation will involve a rise in factor productivity and, hence, other things being equal, improvement in living standards.

How does innovation contribute to development?

In the 'Innovation for Development Report 2010-2011' Report, Augusto López-Claros and Yasmina Mata, in their seminal work entitled 'Policies and Institutions Underpinning Country Innovation: Results from the Innovation Capacity Index', highlighted the historical evolution of the roles of various factors in the development process and the recent role-shift to innovation:

> "*Our understanding of what drives national prosperity has evolved over time. Natural resources, population growth, industrialization, geography, climate, and military might have all played a role in the past. We also know that the relative importance of these drivers has shifted over time, and that in recent decades, more importance has been given to the coherence and quality of policies and the development of supporting institutions. A relative newcomer to this debate –identified as perhaps one of the most important modern engines of productivity and growth –has been the **innovation excellence** of a country; that is, its industries, researchers, developers, creative thinkers, enlightened politicians, managers, and clusters.*" (Lopez-Carlos and Mata, 2010)

Innovation is considered as the engine of social and economic development, in both developed and developing countries. The World Bank's World Development Report (WDR) 1998/99 – Knowledge for Development – asserts that "knowledge is like light. Weightless and intangible, it can easily travel the world, enlightening the lives of people everywhere. Yet billions of people still live in the darkness of poverty –unnecessarily." Though this report's focus was on 'knowledge' and it did not explicitly use the term 'innovation', in the context of this paper, we consider 'knowledge' as an integral part of 'innovation'.

The WDR focused on two important sorts of knowledge and two types of problems that are crucial for developing countries: 'Knowledge about technology' (i.e. technical knowledge or know-how e.g. about farming or health) and 'Knowledge about attributes' (such as product quality, market information, or credibility of a borrower). The Report claims that developing countries have less technical knowledge than industrial countries, and the poor have less than the non-poor. According to the Report, this unequal distribution of know-how across and within countries is termed as 'knowledge gaps'. Developing countries also face myriads of problems related to incomplete 'knowledge of attributes' which the Report termed as 'information problems'. Mechanisms to alleviate information problems, such as product standards, market information, and credit reports are fewer and weaker in developing countries than the developed ones. Information problems and the resulting market failures especially hurt the poor.

What factors determine innovation?

As technology and innovation are becoming major drivers to development process; economic output is no longer mainly a function of capital and labour but, increasingly of innovation and the acquisition of knowledge. Based on these considerations, Lopez-Carlos and Mata (2010) posed certain central issues such as: the factors, policies and institutions conducive for innovation; their relative importance; how they do interact with each other; and how successful countries have been in identifying and adopting them." Some of the key areas identified by Lopez-Carlos and Mata were:

- Education and human capital – levels of spending in education, research and development, and in information and communication technologies.
- Good governance – Is the regulation of labour market appropriate? Or does it provide perverse incentives for workers and employers? Do the government policies encourage the arrival of skilled workers and highly qualified professionals?

- Uncorrupted system (transparency) – Are the public procurement policies and systems open and do they encourage the adoption of new technologies and reward innovation? Are government tax incentives well-targeted and applied transparently or do they distort the incentive system?
- Regulatory framework – Questions such as: What is the legal basis for property (including intellectual and contract design? Where is it easier or more difficult to enforce contracts? Which countries make it easy to get licenses? Where investors are provided the greatest protection? Which countries have the most restrictive labour legislation?
- Does the financial system allow easy access to finance and the emergence of venture capital?
- Does the trade system open and encourage competition?
- What is the degree of collaboration between university and industry? Is the university system delivering to the business community adequately trained graduates?

Innovation capacity index (ICI)

Now a day, it has been increasingly common to construct composite indices in an attempt to measure the performance and trends of various social and economic indicators of countries. The UNDP's Human Development Index (HDI), the inclusive Wealth Index (IWI), the gross national happiness index (GNHI), and the international transparency's corruption perception index (CPI) are some of the examples. In attempt to measure the extent, performance and trend of 'innovation' at a national level, Lopez-Carlos and Mata, have constructed an innovation capacity index (ICI). In constructing the Index, they managed to strike the balance between broad coverage factors that affect a country's capacity for innovation, on the one hand, and a certain degree of economy, on the other, as there may exist potentially large number of variables which could determine a nation's ability to innovate. After identifying a wide range of variables (see appendix 1), they clustered these factors into five categories or pillars (figure 1). These are:

1. Institutional environment
2. Human capital, training and social inclusion
3. Regulatory and legal framework

4. Research and development
5. Adoption and use of information and communication technologies

Lopez-Carlos and Mata have ranked countries on the basis of the ICI. In 2010-2011, they constructed ICI for 131 countries. In this ranking Sweden received the 1st the rank and Ethiopia the 108th place out of the 131 countries considered in ICI ranking, among the lowest in ICI.

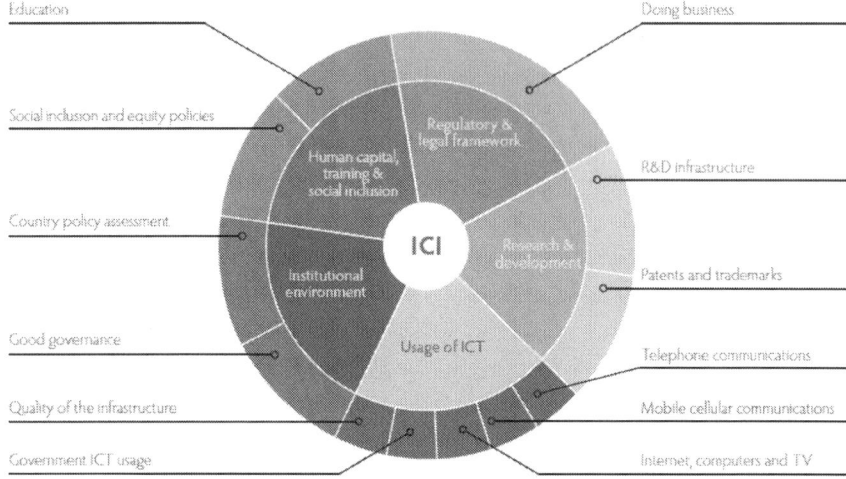

Figure 1. The innovation capacity index
(Source: Lopez-Carlos and Mata (2010), p.18. In: the Innovation for Development Report 2010-2011).

'Innovation' – the 'heart' of Schumpeterian economics

Joseph Alois Schumpeter (1883 – 1950) was an Austrian-American economist and political scientist. He was one of the most influential economists of the 20th century. Schumpeter popularized the term "creative destruction" in economics. In the Schumpeterian economic theory, innovation is closely related to development – economic development is driven by the discontinuous emergence of new combinations (innovations) that are economically more viable than the old way of doing things (Schumpeter, 1934). Schumpeter's innovation concept covers five areas:

1. the introduction of a new good or a new quality of a good (product innovation);
2. the introduction of a new method of production, including a new way of handling a commodity commercially (process innovation);

3. the opening of a new market (market innovation);
4. the conquest of a new source of supply of raw material or intermediate input (input innovation); and
5. the carrying out of a new organisation of industry (organisational innovation).

According to Schumpeter, it is the introduction of new product, new way of doing, and the continual improvements in the existing ones that lead to growth and development. Schumpeter says that 'entrepreneur' is such a factor of production who introduces new combinations of factors of production. An entrepreneur is neither a technician nor he is a finance manager. Entrepreneurs make innovations just for its own sake or influenced by the desire of profit and socio-cultural set-up of the society. In order to perform this economic function the entrepreneur needs two things: (i) technical knowledge so that he could produce new goods and (ii) access to funds/credit.

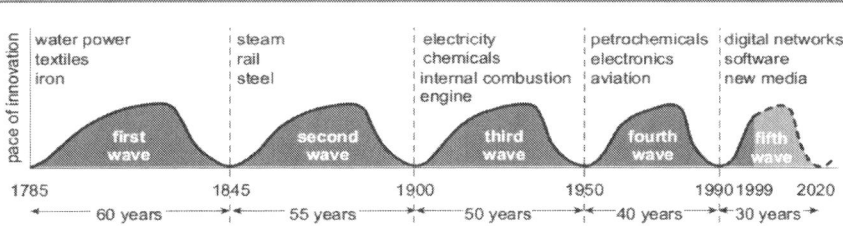

Figure 1 Schumpeter's waves accelerate

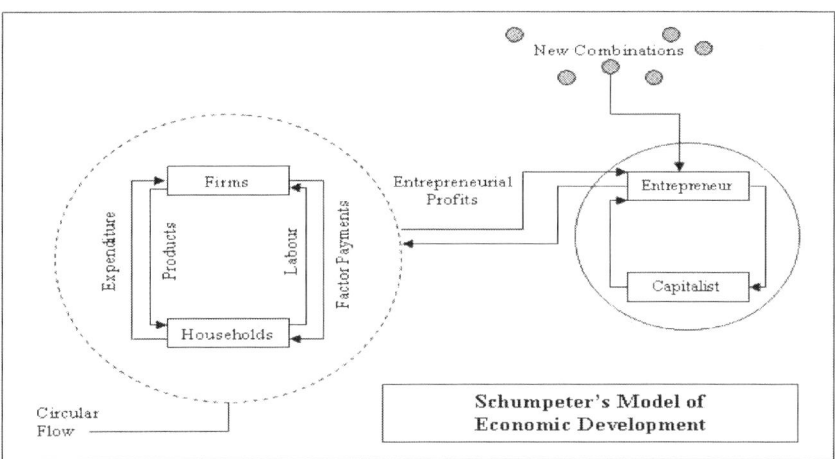

Figure 2. Diagrammatic representation Schumpeterian economic development
Source: A. Pal (n.d.)

The Schumpeterian model of economic growth moves round the 'webs' of inventions and innovations (figure 1). And the actual Schumpeterian model of economic development is represented in figure 2. In Schumpeterian theory innovation is conceptualized as the breaking up of old combination/ old order with rising profit. High profits attract imitators who eventually spread the new way of doing things and create a new order. In both the new and old economic order there is a low profit level which ultimately leaves a way for another web of innovation.

A tale of innovation from Ethiopia - Bethlehem Tilahun Alemu – innovator (shoemaker from recycled materials)

- Motivation (Background)- Bethlehem observed youth unemployment in the area (poor community) despite their talents; attracted by the hard work of her parents; observed the idea of making things by hand and using local materials by local people were there in Ethiopia for long time; observed potential foreign demand for local products; and observed the potential of 'innovative' venture to change local economic conditions;
- Established the company, SoleRebels, nine years ago (ca. 2004).
- Started the company in Zenebe-work area, the poor community in the outskirts of Addis Ababa where she was born with an investment of less than $10,000 (£6,400), put together by her immediate family.
- In 2012 she had 75 full-time employees in the factory and more than 200 local suppliers of raw materials.
- Introduced modern design for local products and take into account the trends followed by consumers in the West.
- The Factory produces around 800 pairs of shoes a day which are sold at a price of - on average - between $35 and $95.
- Now, one of Ethiopia's most thriving businesses. The company sells its products in 55 countries, (its biggest markets are in Austria, Canada, Japan, Switzerland and the United States); also sells online.
- She uses old tyres, natural fibres and hand-made fabrics - all locally sourced - to manufacture sandals and other shoes which are inspired in the traditional Selate and Barabasso tyre footwear once worn by Ethiopian rebels.
- International Prizes:

- the planet's first fair trade green footwear firm - certified by the World Fair Trade Organization (WFTO).
- selected as a Young Global Leader by the World Economic Forum in Davos, Switzerland (2011),
- One of the winners of the Africa Awards for Entrepreneurship in Nairobi, Kenya.
- Bethlehem was listed by the US business magazine Forbes as one of Africa's most successful women (2012).
- She received the Social Entrepreneur of the Year Award at the 2012 World Economic Forum on Africa.

- She now plans to build a bigger manufacturing plant where she hopes to employ up to 300 people. "We are doing well. We are trying to do $2m this year. In 2016, we are planning to do $20m. So that's why we are working hard and we are trying to expand our working facility," Mrs Bethlehem said.

Source: http://www.bbc.co.uk/news/world-africa-18998898

3.
CONCLUSION AND POLICY IMPLICATION

Developing countries such as Ethiopia are facing myriads of challenges – economic, social, governance, and environmental. From the experiences of development paths of many developed and developing countries it can be safely generalized that there is no a 'one-fit-for-all' remedial 'silver bullet' for development. Despite this fact, it has been increasingly evident, particularly in the 21st century, that pursuing 'old way' of doing things may unlikely leads a nation to the destiny of development. Citizens should be able to be innovative – think and act differently. The tale of Bethlehem in Ethiopia illustrates the huge untapped potential for innovation and development in Ethiopia. Yes, Schumpeterian Economics has a place in Ethiopia! However, there are numerous barriers to innovation in Ethiopia – restrictive policies, lack of good governance, corruption, backward technology (including ICT) etc. From the citizens' side, lots of young people are developing the culture of a 'windfall gain' and very biased view and growing belief/tendency that one has to move abroad in order to 'win poverty'. This tendency needs to be averted.

What should be done in Ethiopia to promote innovation? Well, government should understand the underlying constraints to innovations and attempt to find solutions to problems and create conducive environment to innovation. This may include, (a) develop pro-innovation policies and promote new ideas and thoughts, (b) remove certain restrictive policies and allow the free flow of information, (c) support cooperation, production and sharing of information, (d) gear education and research to provide evidence-based case studies of good practice, success stories and lessons learnt, and the factors promoting and impeding innovation, (e) human capacity-building to promote innovation, (f) promotion of a wider awareness and public understanding of innovation at the practitioner, programme and policy levels to ensure innovation gets effectively onto the Development Agenda.

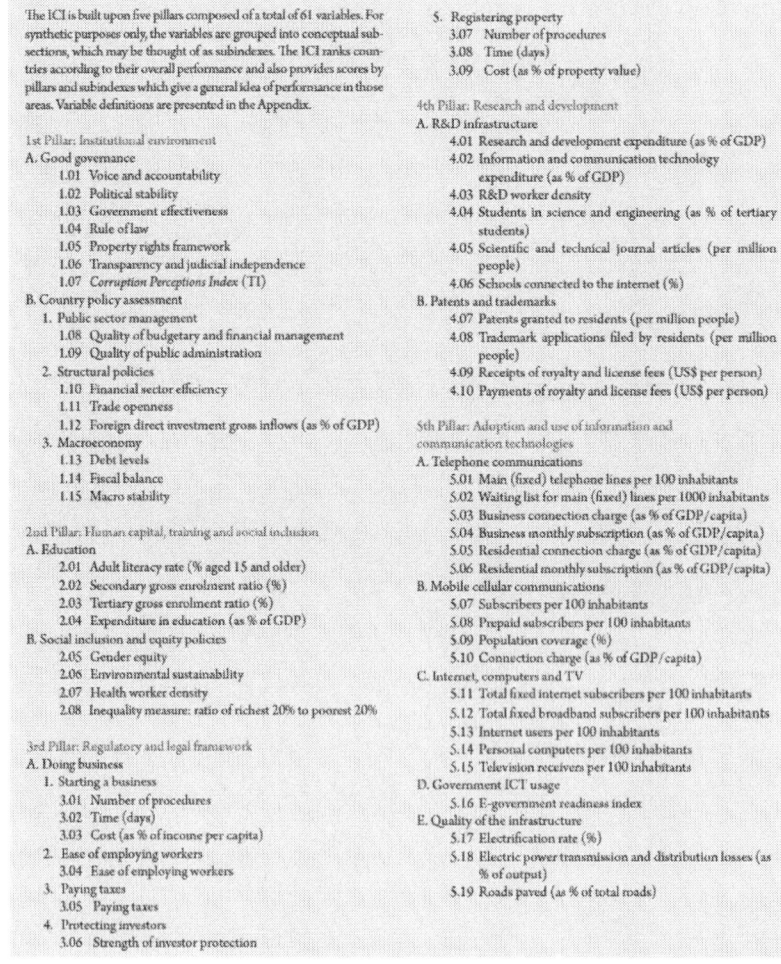

Appendix 1. Structure of the innovation capacity index
(Source: Lopez-Carlos and Mata (2010), p.18. In: the Innovation for Development Report 2010-2011).

REFERENCES

BBC: http://www.bbc.co.uk/news/world-africa-18998898 [Accessed 17 November 2013].

Lopez-Carlos, A., Mata, Y.M., (2010). Policies and Institutions Underpinning Country Innovation: Results from the Innovation Capacity Index. In: The Innovation for Development Report 2010-2011. Microsoft Corporation.

OECD and European Communities. (2005). Oslo Manual: Guidelines for Collecting and Interpreting Innovation data. Joint Publication of the OECD and the Statistical Office of the European Communities.

Onodera, Osamu and Earl Kim Hann, (2008). Case Study 2: Trade and Innovation in the Korean Information and Communications Technology Sector. OECD, Paris.

Schumpeter, J.A., (1934). The Theory of Economic Development: An Inquiry into Profits, Capital, Credit, Interest and the Business Cycle. Cambridge, MA, Harvard University Press..

Schumpeter, J.A., 1934. The Theory of Economic Development: An Inquiry into Profits, Capital, Credit, Interest and the Business Cycle. Cambridge, MA, Harvard University Press.

The World Bank, (1999). Knowledge for development. World Development Report 1998/99. New York, Oxford University Press Inc..

The World Bank, (2001). Building institutions for markets. World Development Report 2002. New York, Oxford University Press Inc.

Todaro, M.P., (1998). Economic Development (6th ed.) USA, Addisob-Wesley Longman Ltd.

UNESCO, (2009). Innovation for Development – converting knowledge to value. OECD-UNESCO international workshop Summary Report. Paris.

CHAPTER 5:
A Conceptual Framework for Building a Homegrown Public-Private Partnership Platform to Deliver Public Information Services in Developing Economies

Temesgen A. Weseni and Richard T. Watson

1.
INTRODUCTION

How can academicians and practitioners accurately evaluate the role of partnerships for the success of Information Systems (IS) as well as for the delivery of public information services? Answering such multifaceted question is not that easy. Traditionally, the role of IS has been to design, build, and install systems to improve organizational performances but today IS needs to look beyond systems building (Watson et al., 1998). Following such strong comments, there have been limited amounts of IS studies attempting to define and theorize the success of public information services delivery and these efforts have been partly satisfactory or totally ignored the role of partnerships. Public information services delivery literally refers to the use of technology, especially web-based applications to enhance access to and efficiently deliver government information and services (Brown and Brudney, 2001). It incorporates an implementation of cost effective models for citizens, industry, government employees, and other stakeholders (Whitson and Davis, 2001).

It is a use of Information Communication Technology (ICT) as a tool to achieve better government (OECD, 2003) which subsequently electronic services have become governments' priority (Harfouche and Kalika, 2009).

Nowadays governments are no longer considered the sole provider of public works and services because of the forces driving this movement such as scarcity of public resources, a political trend toward the deregulation of infrastructure, and an expansion of global markets (Ababutain, 2002). In this regard, the involvement of the private sector could also be a means of introducing and transferring new technology which is especially important in developing economies (Blaiklock, 2003). Partnerships in which public services are provided using private infrastructure are increasingly common in low and lower-middle income economies where many people cannot afford or do not have access to the Internet (UNPAN, 2012). As the private sector is less concerned in equity and transparency than is the public sector (Rosenaue, 1999), partnership with the public would allow some risks to be transferred to the private sector and hence to the parties best able to manage projects which will result in gains in performance and productivity (Zou et al., 2008). To cope with such gradual trends, nowadays governments are focusing on Public Private Partnerships (PPPs). Although, PPP is not easy to apply (Boeva and Vassileva, 2008), development of PPP is an alternative method of implementing public sector infrastructure projects as part of government's role of promoting sustainable economic development where government allows the participation of private sector in developing and implementing an infrastructure business through carefully integrating environmental, economic, and social needs to achieve both an increased standard of living in the short term, and net gain among future generations (Rashed et. al., 2011).

In essence, PPP approach can have a strong positive effect on the economic life of any country (Montanheiro, 2008). The effect of synergy (in partnership) can be put into a metaphorical formula: 1 (public) + 1 (private) > 2 (Wang, 2009). A successful partnership between the public and private sectors depends on all of the people involved with the project (NASCIO, 2006). There are three main reasons for adopting the PPP approach. Firstly, the private sector possesses better mobility than the public sector and therefore the private sector is not only able to save the costs of project in planning, design, construction and operation, but also avoid the bureaucracy and to relieve the administrative burden. Secondly, there is a wide spread belief that the private sector can provide better service to the public sector and establish a good public private partnership so that balance risk-return structure can be maintained. And lastly, governments' inability to raise massive funds for large-scale infrastructure projects can be mitigated by private participation (Cheung et al., 2009).

In conclusion, delivering services through PPP is utilized most in Europe and Asia, 56 and 53 per cent, respectively, however, significantly the lowest in Africa which is only 17 per cent (UNPAN, 2012). Unfortunately, the technology often is the "scapegoat" within an unsuccessful partnership (NASCIO, 2006). Beyond these reasons, in the case of PPPs, over the past three decades, governments in both developed and developing economies have embraced PPP as an alternative to the standard models of public procurement strategy to deliver public services and this is especially true for governments lacking public sector resources to deliver important public services (Rashed et. al., 2011).

2.
PUBLIC PRIVATE PARTNERSHIP APPROACHES

In a broader terms, although Public–Private Partnerships (PPPs) can bring greater efficiency in the delivery of services, PPPs cannot replace the public sector or regulator nor operate without reasonable profit for private sector and sustainability for the public, nor exist in the absence of political will in a country (The Asia Foundation, 2010). PPP, in its essence, have long been advocated and analysed as organizational solutions to pressing societal problems that call for the comparative advantages of government, business, and civil society, however, ongoing questions remain about how to design, manage, and assess PPPs (Derick and Jennifer, 2011).

Nowadays PPPs cover wide areas of cooperation and arrangement between governments and privates sectors, local communities and others stakeholders on the matter of dealing public information service deliveries. Gallegos (2012) identified various options of PPPs that allocate responsibilities and risk of operation between public and private sectors and other concerned stakeholders in the game to play properly the role while delivering the services. The report incorporates the following as main models: (i) BOT (build-operate-transfer), (ii) BOOT (build-own-operate-transfer) (iii) BOO (build-own-operate) (iv) BTO (build-transfer-operate) (v) LDO (lease-develop-operate) (vi) ROT (rehabilitate-operate-transfer) (vii) DBFO (design-build-finance-operate) (viii) IPO (Initial-Public-Offering), and others service contracts.

ESCAP (2011), on the other hand, listed out the conventional criteria of classifying the approaches to PPP which include (i) ownership of capital assets; (ii) responsibility for investment; (iii) assumption of risks; (iv) duration of contract, etc. Taking these standards in to account, PPP models can be classified into five broad categories: (i) supply and management contracts, (ii) turnkey contracts, (iii) Lease, (iv) concessions, (v) private finance initiative and private ownership.

The major benefits of PPPs for the general public and its governments are the delivery of instant, efficient and effective information services associated with private business practices to public service in order to ensure and satisfy citizens. In this regard, the involvement of the private sector in public services delivery also supports the business process to be transparent and competitive; as a result, the long-term costs of the service delivery can be assessed more realistically under a PPP framework which in turn promotes efficient use of resources (Hussen, 2013). Due to these factors, PPP is increasingly being seen as an answer to several challenging problems that the public agencies in general face in serving their population effectively. This is especially true in developing economies, where generally the public agencies face resource constraints in improving their operations and delivering services to their needy citizens.

3.
PARTNERSHIP CHALLENGES IN DELIVERING PUBLIC INFORMATION SERVICES IN ETHIOPIA

Africans created the world's first major information systems, gesturing and language (Watson, 2013). However, solutions to Africa's delivery of information service challenges require the combined efforts of the private, public, and voluntary actors of various sectors. The question is, while this issue observed by many African countries, how best to combine those efforts continue to be a topic of ongoing debates.

Poverty reduction, improving civil service systems and transformation of public services are among the current hot issues of low-income countries. Ethiopia, one of the developing economies, is striving to fulfil the standards set under the United Nation's Millennium Development Goals (MDG) and specifically its own five year Growth and Transformation Plan (GTP). In such a densely populated country with scarce resources and lack of know-how, in order to address challenging issues related to quality of public service information systems, the Government of Ethiopia is currently giving due attention and concern by considering PPP strategies for its public services. The Ministry of Communication and Information Technology (MCIT) clearly recognizes the role of PPPs in delivering public information services by establishing relationships with the private sector ICT firms (PWC, 2011).

Channels of service delivery, in this regard, are the ways of communication through which a service is delivered to the citizen (Sousa and Voss, 2006). It is also the way by which a citizen requests a public service and receives the resultant output from a service (Harfouche and Kalika, 2009). Services as a series of interactions between the service provider and clients that result in an observable output (Janseen et al., 2009). As such, ICT is being viewed as a key tool to bring about a change in service delivery approaches (UNDESA, 2008). Among the potential values and efficient uses of ICTs is that to deliver improved services to citizens (Coleman, 2006).In this regard, the use of ICT would be to promote more efficient and effective government, facilitate the accessibility of government services, allow greater public access to information, and make governments more accountable to citizens (Kitaw, 2006). In this regard, developing economies, however, that couldn't effectively manage their information resources mainly because of its public system failed to work with private partners and fill the gaps of basic public information services as a result it would persist on that of conventional manual and poor civil service systems.

Following the gradually growing demand for the acceleration of infrastructural development and the improvement of service delivery in Ethiopia, considerations of PPPs as a model for public services development and information delivery is increasing. Although various underlying benefits of such arrangement have been enumerated and continue under the consideration of the Ethiopian government, there exist challenges in managing and selecting the right PPPs which manifest in a number of unobserved failure or success cases. In line with this, the primary possible barrier that could be identified therefore is the level of caution within the public sector as clearly argued by Gunnigan and Eaton (2008), there is political pressure to ensure that PPP projects do not compare unfavourably to the traditional projects and cost to the taxpayer will be a factor in political debates.

The Ethiopian Government has recognized the power of ICT in the national development plan and this is indicated by ratification of the National ICT Policy and setup new Intuition at a Ministry level to lead the sector, as well as allocating sufficient resources for ICT development. Since the Ethiopian "e-Government Strategy and Implementation Plan" released in 2011, the government has started to build up major public services with a collaboration of the private sectors using PPP strategies for selected critical e-Services enablement. According to the plan, through the implementation of twelve Agencies' priority projects and through four alternate channels of services delivery, the State expected to facilitate the creation of a sustainable IS ecosystem (PWC, 2011).

As far as the existing Ethiopian public information service delivery is concerned, the presence and background of basic civil services in most parts of this country exist as manual and haphazard. Even if there is a growing recognition of the role of PPPs for the majority of Ethiopian public sector information services in general, the current progress and status of different agencies and utility service providers appears to be lacking proper experience and knowledge of initiating IS PPPs. Although the business climate of the country from the first glance looks like an attractive spot for investing and establishing partnerships however the readiness of the public sector for establishing partnerships with the private sector seems on a premature and underdeveloped level. Though some exemplar initiation of PPPs are ongoing such as the current new service called "*Lehulu*" which is a network of centres providing a Unified Billing System that allows one to pay all utility bills (Electricity, Water, and Landline phone) by merging the previous three public services' payment into one-window, still poor qualities of services of IS are observable in many parts of the public sector of the country.

Above all the telecom and other IT-related industries are found unquestionably in their infant stage of development for establishing partnerships due to known and unknown factors and thus calls for study which aims to promote the knowledge of such PPP activities in major critical sectors of the country and to come up with valuable outputs. In a nutshell, with all such PPP efforts the Ethiopian e-Services are still not as strong as expected. For example, according to the survey of e-government development index in year 2012 of the United Nations, Ethiopia is ranked at 172nd place (UNPAN, 2012). This is relatively low score even among some other African countries such as Uganda, Rwanda, and Malawi. Needless to say, keeping in mind the above challenges, the motivation of the researchers is regarded from the above stated points and the following one multi-faceted question longing to be answered: what should be done in order to utilize PPP in delivery of information services in Ethiopia?

4.
GUIDING THEORETICAL FRAMEWORKS FOR THE RESEARCH

Researchers have proposed different theoretical frameworks to understand the reasons for the failure of numerous public sectors' initiatives of electronic government developments especially the case of ICT initiatives of the developing economies. Although the role of PPPs initiatives for successful delivering of public information services is becoming one of the appreciated issues for many IS researchers, but there lack a comprehensive model in IS literatures that help to assess the issues comprehensively. This study, on the other hand, mainly utilizes the 'Stakeholder Theory' (ST) and Actor-Network Theory (ANT) to explore the features of PPP initiatives in delivering public information services' success and challenges in the developing economies specifically in the context of Ethiopia. In general, from these two theoretical views one can potentially add paramount contributions to the study.

Stakeholder Theory

Various literatures on Stakeholder Theory and its analysis shows that several authors have attempted to identify as well as classify stakeholders and their role. Clarkson (1995) classifies them as the primary stakeholders, who are essential to survival and wellbeing of the organization, and the secondary stakeholders, with who an organization interacts, but the interactions are rather complementary than essential. Mitchell et al. (1997) also provide a comprehensive framework that explains how managers prioritize stakeholders' relationships by identifying (i) stakeholders' power to influence a firm, (ii) the legitimacy of the stakeholders' relationships with a firm, and (iii) the urgency of the stakeholders' claims on a firm.

Some authors also suggested that measuring the level of influence of stakeholders and the importance of their relationships by first identifying the stakeholders, then by categorizing them, and eventually connecting them with different types of arrays for example Carroll and Buchholtz (2003) applied the 'STEP' model with its four major contributing environments, i.e., social, technological, economic and political, to identify the stakeholders (their sub-elements) and their relationships and influences. Clulow (2005) also proposes a systematic discourse analysis that goes beyond the identification of key stakeholders, and different perspectives, including economy, sustainability and responsibilities, should be considered.

Bourne and Walker (2005) have provided a mapping tool to visualize stakeholders' power, influence and contribution within the performing organization. Enserink et al. (2010) also suggest to follow a six-step stakeholders analysis (1) problem formulation, (2) inventory of the stakeholders involved, (3) development of a chart to illustrate stakeholders formal tasks, authorities, relations and current legislation, (4) determining the interests, objectives, and problems, (5) mapping out the interdependencies between stakeholders, and eventually (6) determining the consequences of these findings with regard to the initial formulated problems. Hummel et al. (2004) affirm that the use of methods supporting and managing the knowledge from involved stakeholders improves the processes of design of stakeholders' model. Bergman et al. (2007) also emphasize the relevance of the organizational and political context surrounding design. Their model describes that significant improvements in systems design can be achieved by focusing on questions, (1) what system(s) can be built and delivered within the given environment, and (2) how to align stakeholders' interests with the proposed designs to mobilize willingness and resources. Generally, one can argue that these great contributions of Stakeholder Theory and analysis approaches are should be considered throughout the PPP projects in order to deliver public information services because as Bouwman et al. (2010) discussed that stakeholder analysis should not serve as a validation of the business/service concept. It rather should capture and evaluate the dynamic behaviour and interests of stakeholders during the innovation/design project continuously, up till the final phases of the project.

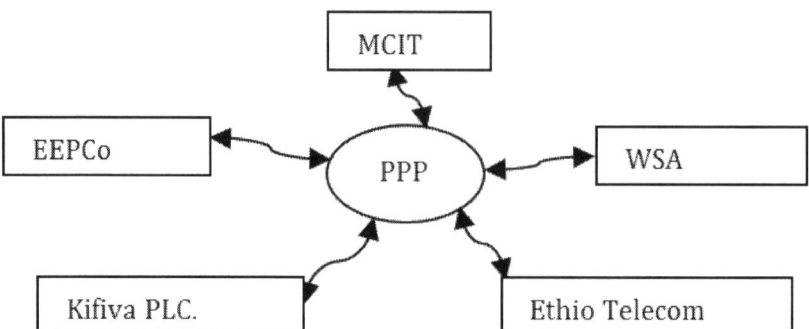

Figure 1: Stakeholders Involved in the Ethiopian PPP (spider-view)
[Researcher's Preliminary Investigation]

Actor-Network Theory (ANT)

Actor-Network Theory (ANT) declares that the world is full of hybrid entities containing both human and non-human elements, and was developed to analyse situations where separation of these elements is difficult (Callon, 1997). ANT also examines the motivations and actions of actors who form elements, linked by associations, of heterogeneous networks of aligned interests (Walsham, 1995). To comprehend complex IS development processes ANT researchers do not take the social as given but instead follow the actors as they enrol into heterogeneous actor networks and thus assemble the social. By employing ANT many IS researchers provided robust accounts of the production and reproduction of actor networks in the development and implementation of IS thus enabling deeper understanding of their failure or success (see for e.g. Underwood, 1998; Walsham and Sahay, 2006).

As Underwood (1998) explains:

> *Actor-network theory helps us to understand the course of a project or enterprise. We can ask questions such as "How did it come to turn out this way?" (through the changing alliances of actors), "Who is influencing it?" (who has been doing what scripting?) or "Why are some actors acting this way?" (what scripts are they carrying?). These are not questions with deterministic answers but they allow a rich interpretation of the situation.*

As ANT has capability of providing insights into socio-technical settings where human and non-human agents interact, it is going to be used its concepts and contributions as a theoretical lens for this research with the objective of identifying main factors related to the participated actors that led to the PPP initiatives more localized success and by taking into account main information such as: the actors' gains, efforts, and relationships.

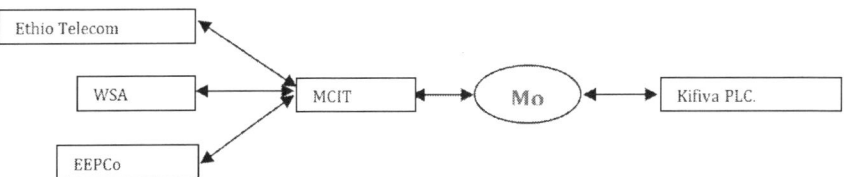

Figure 2: A Networking View of Actors in Ethiopian PPP
[Researcher's Preliminary Investigation]

5.
RESEARCH DESIGN AND METHODOLOGY

The strategy selected for this research is an in-depth case study of a single PPP case in which this specific chosen case has exactly incorporated a total of four Ethiopian government firms and a single private firm. Given the interpretive stance adopted for this research and considering of the nature of its one multifaceted research question, a case study approach happens to be an appropriate research strategy for this study. Single case studies allow researchers to investigate phenomena in depth to provide rich description and understanding (Walsham, 1995). In fact, Yin (1994) has also endorses and explaining that a single case can often produce a more penetrating study.

Because of case study largely relies more on studying in-depth and its hallmark is the use of multiple data sources which enhances data credibility (Yin, 2003), we will focus on one actively operational PPP case which existed for the first time in Ethiopia under the category of CSCs (Common Service Centres) in the "*e-Government Strategy and Implementation Plan*" of the 2011 MCIT's (Ministry of Communication & IT) plan. The plan clearly state that the government is starting to build up major public information service delivery channels with a collaboration of the private sectors using PPP strategies for specially selected critical services implementation of which the State is expected to facilitate the creation of sustainable IS ecosystem (PWC, 2011).

CSC type of PPP incorporates services like information dissemination, acceptance of public service requests and delivery of information services provided to the customers at a single point of public service delivery. CSC in general includes self-service kiosks, utility bill payment centres, ICT community centres etc. According to the MCIT's plan, it is estimated that an investment of approximately 2206.21 Million Birr would be needed over a period of five years for additional 800 CSCs to be set up in partnership with the private sector on a PPP business model which will cover all the Woredas by spreading down till Kebele level based on need and demands of different Regions of the country (PWC, 2011).

Specifically this research will focus on one of the CSC PPPs which initiated and already started providing its service to the general public of Ethiopia as single window based UBS (Unified Billing System) for utility payments. This PPP business model started operating for the first time in the country by integrating one local private firm namely Kifiya Financial PLC and four public stakeholders namely Ethio-Telecom, WSA (Water & Sewerage Authority), EPCo (Electric Power Corporation), and MCIT (Ministry of Communication & IT). This UBS is officially called *"Lehulu"* (Literally in English *"For-All"*) which started operating in 2013 by opening more than 33 operating branches in the capital and at the same time having a plan to commence shortly by expanding its services in four major cities of the country namely Mekele, Bahirdar, Hawassa, and Diredawa. Essentially, this PPP project has also planned to expand and establish in total up to 70 billing centres (for payment of power, telephone and water bills) in the country by fulfilling with all required IT Infrastructure (PWC, 2011).

There are several reasons for choosing this PPP case specifically. Firstly, although the Ethiopian economy is perceived to be one of the leading African emerging markets with double digit growth, the country's major public information service delivery channels are however widely recognized as being seriously haphazard and underdeveloped. Secondly, public sectors such as MCIT nowadays are giving special attention to PPP business models and are starting to motivate the private firms to integrate and initiate to work with them (PWC, 2011). Thirdly, private IT firms such as Kifiya Financial PLC performing a tremendous amount of efforts and information-intensive activities to alter the well-aged poor delivery of public information service including UBS.

Fourthly, the undelivered and neglected public information services touches on the daily lives of a broad section of developing economy's citizens. These miserable people who are most dependent on the public sector's services remain one of the biggest challenges for the public system itself. Thus the role of such PPP is going to be enormous. Last but not least, the researcher's own personal involvement in such issues led him to observe and learn the role of PPPs' contribution not only to the Ethiopian context but also to the overall development of public information service of the developing economies.

Research Approach, Data Collection Techniques, and Analysis

Qualitative research is more suitable for studies that are rich in detailed descriptions around context and processes (Kaplan and Maxwell, 1994). Accordingly, in order to answer the main multifaceted research question of this study, a qualitative approach will be employed to gather data from oral discourses. Moreover, qualitative research also used for studying selected issues, cases or events in depth and in details (Orodho and Kombo, 2002). This approach, in general, will expect to give greater scope to address the research issues and to ask questions such as 'why' and 'how' particular trajectories are created. Furthermore, since the units of analysis are concepts, the aim is mainly to construct a theoretical explanation by specifying the conditions and processes in a phenomenon.

Both primary and secondary data will be collected in relation to delivery of public information service projects under a PPP alliance from four governmental stakeholders namely Ethio-Telecom, WSA (Water & Sewerage Authority), EPCo (Electric Power Corporation), and MCIT (Ministry of Communication & IT) and one local private firm namely Kifiya Financial PLC. This data will come from detailed interviews with the private and government officials involved in PPPs at different levels of decision making, project participants, service operators, and users of the services who are found in Addis Ababa City. Similarly, focus group discussion will be made with the beneficiaries of the public services about their experiences and perceptions towards the result of such partnerships. Specifically, interviews and participant observations will be serving as main sources of data. All in-depth interviews will be supplemented by document review such as using memoranda, organizational charts, project documentation, and related documents such as annual reports, manuals, minutes, and newsletters will be consulted. By considering the aforementioned sources, the researchers will thus be able to triangulate findings.

6.
POTENTIAL CONTRIBUTION OF THE STUDY

PPP is mainly aimed at forming an alliance between actors and empowering the traditional and haphazard public services by improving its overall efficiency and effectiveness through the use of the privates sector's expertise and technology into the public sector's capacity building program. Extraordinary investments are made on PPP projects in spite of the limited financial resources and other constraints of the developing world including Ethiopia. It is, however, questionable why IS literatures or IS researchers neglect investigating such valuable PPPs initiatives regarding failure or success. This research is therefore aimed at developing a conceptual framework to assure of such PPP initiatives. The research, on one hand, aims to contribute to the theoretical domain of IS in general and PPP in particular by adding new insights into the literature by uncovering contributing factors of those PPP initiatives. All selected theories will be consulted in order to devise a comprehensive conceptual framework that can nurture PPP initiatives in developing economies in a better way. The outcome of this research can also be used as a spring-board for future theoretical deliberation by other researchers engaged in PPP researches.

The study is also aimed at developing practical implications for policy makers and practitioners who are engaged in assessing possibilities of PPPs, attracting partners, introduce PPPs, implement PPPs, manage PPPs, monitor PPPs as well as evaluate initiatives of PPPs in Ethiopian context. As PPP is considered as a driver and a major enabler for rapid development in a country, the output of this research can be used as important input to the policy makers and other concerned governmental bodies in their effort to make the existing public services more efficient and effective. It can also be used as valuable input to assess PPP initiatives and to devise the way how future PPP efforts as well as future public information services can be approached by practitioners.

REFERENCES

Ababutain, A.Y. 2002. A Multi-Criteria Decision-Making Model for Selection of BOT Toll Road Proposals within the Public Sector. PhD Thesis, Civil and Environmental Engineering Department, University of Pittsburgh.

Bergman, M., Lyytinen, K., and Mark, G. 2007. Boundary Objects in Design: An Ecological View of Design Artefacts. *JAIS,* 8(11), 546-568.

Blaiklock M. 2003. An Introduction to Public Private Partnerships: A Review of the Key Issues, European Construction Institute, Loughborough University, Loughborough. pp. 7-12.

Boeva, B. and Vassileva, A. 2008. Public-Private Partnerships: Formulae for Success, Incentives and Barriers. Economic Alternatives, Issue 2.

Bourne, L. and Walker, D.H.T. 2005. Visualizing and Mapping Stakeholder Influence, *Management Decision,* 43(5), 649-660.

Bouwman, H., de Reuver, M., Hampe, F., Carlsson, C., and Walden, P. 2010. Mobile R&D Prototypes: What is Hampering Market Implementation? 9th International Conference on Mobile Business, Ninth Global Mobility Roundtable, pp.17-24.

Brown, M.M. and Brudney, J.L. (2001). Achieving Advanced Electronic Government Services: An Examination of Obstacles and Implications from an International Perspective. Paper Presented at the National Public Management Research Conference, Bloomington.

Callon, M. (1997). 'Actor-Network Theory - The Market Test' Actor Network and After Workshop. Centre for Social Theory and Technology (CSTT), Keele University, UK.

Carroll, A. and Buchholtz, A. (2003). Business and Society: Ethics and Stakeholder Management, 5th ed., Thomson South-Western, Mason, OH.

Cheung, E., Chan, A., and Kajewski, S. (2009). Reasons for Implementing Public-Private Partnership Projects: Perspectives from Hong Kong, Australian and British Practitioners. *Journal of Property Investment and Finance,* 27(1), 81-95.

Clarkson, M.B. 1995. A Stakeholder Framework for Analysing and Evaluating Corporate Social Performance. *Academy of Management Review,* 20(1), 92-117.

Clulow, V. (2005). Future Dilemmas for Marketers: Can Stakeholder Analysis Add Value? *European Journal of Marketing,* 39(9-10), 978-997.

Coleman, S. (2006). African e-Governance – Opportunities and Challenges, University of Oxford, Oxford University Press.

Derick, W.B. and Jennifer M.B. (2011). Public–Private Partnerships: Perspectives on Purposes, Publicness, and Good Governance. *Public Administration and Development,* 31(1), 2-14.

Enserink, B., Hermans, L., Kwakkel, J., Thissen, W., Koppenjan, J., and Bots, P. (2010). Policy Analysis of Multi-Actor Systems. Lemma, The Hague.

ESCAP. (2011). A Guidebook on Public-Private Partnership in Infrastructure. Economic and Social Commission for Asia and the Pacific (ESCAP). United Nations.

Gallegos, D. (2012). 'Partnerships for Broadband: Innovative Public Private Partnerships Will Support the Expansion of Broadband Networks.' ICT Policy Notes, Note 2. Washington, DC: World Bank.

Gunnigan, L. and Eaton, D. (2008). Barriers to Innovation in Public-Private Partnership (PPP). Proceedings of COBRA–Dublin Institute of Technology.

Harfouche, A. and Kalika, M. (2009). E-Government Implementation and Acceptance: Challenges to Increase Public e-services Take-Up in Lebanon. AMCIS 2009 Proceedings.

Hummel, M., van Rossum, W., Verkerke, G.J., and Rakhorst, G. (2002). Product Design Planning With the Analytic Hierarchy Process in Inter-Organizational Networks. *R&D Management*, 32(5), 451- 458.

Hussen, M. 2013. Public-Private Partnerships (PPPs) as Collective Action for Improved Urban Environment Governance in Ethiopia. *Journal of Economics and Sustainable Development*, 4(10), 59-66.

Janssen, M., Chun, S.A. and Gil-Garcia, J.R. (2009). "Building the Next Generation of Digital Government Infrastructures," *Government Information Quarterly*, 26(2), 233-237.

Kaplan, B. and Maxwell, J.A. (1994). Qualitative Research Methods for Evaluating Computer Information Systems. In Evaluating Health Care Information Systems: Methods and Applications, J.G. Anderson, C.E. Aydin, and S. J. Jay (eds.), CA: Sage, p.45-68.

Kitaw, Y. (2006). E-Government in Africa: Prospects, Challenges and Practices; International Telecommunication Union.

Mitchell, R., Agle, B., and Wood, D. 1997. Towards a Theory of Stakeholder Identification and Salience: Defining the Principle of Who and What Really Counts, *Academy of Management Review*, 22(4), 853-886.

Montanheiro, L. 2008. Public-Private Partnerships and their Economic Contribution. *International Journal of Applied Public-Private Partnerships*, 1(2), 1-18.

NASCIO. (2006). Keys to Collaboration: Building Effective Public-Private Partnerships. The National Association of State Chief Information Officers (NASCIO). Corporate Leadership Council (CLC).

OECD. (2003). Background Paper: Implementing e-Government in OECD Countries: Experiences and Challenges. OECD-OCDE.

Orodho, A. J. and Kombo, D.K. (2002). Research Methods. Nairobi: Kenyatta University, Institute of Open Learning.

PWC. (2011). E-Government Strategy and Implementation Plan. MCIT.

Rashed, M., Alam, M., and Toriman, M. 2011. Considerable Issues for Sustainable Public-Private Partnership Project. *Scientia journals, Res Manageria*, 2(4), 57-65.

Rosenaue, V. (1999). The Strengths and Weaknesses of Public-Private Policy Partnerships. *American Behavioural Scientist*, 43(1), 10-34.

Sousa, R. and Voss, C. (2006). Service Quality in Multichannel Services Employing Virtual Channels, *Journal of Service Research,* 8 (4); 356-371.

The Asia Foundation. (2010). Promoting Public-Private Partnership in Bangladesh. Dhaka: Rainbow Enterprise.

Underwood, J. (1998). Not another methodology: what ANT tells us about systems development, The 6th International Conference on Information Systems Methodologies. British Computer Society.

UNDESA. 2008. World Population Prospects. United Nations, Department of Economic and Social Affairs. UN.

UNPAN. 2012. United Nations E-Government Survey-2012. Department of Economic and Social Affairs. New York: The United Nations.

Walsham, G. and Sahay, S. (2006). Research on Information Systems in Developing Countries: Current Landscape and Future Prospects. *Information Technology for Development,* 12 (1), 7–24.

Walsham, G. (1995). Interpretive Case Studies in IS Research: Nature and Method, *European Journal of Information Systems,* 4(2), 74-81.

Wang, Y. (2009). A Broken Fantasy of Public–Private Partnerships. *Public Administration Review.* 69(4), 779-782.

Watson, Richard T. (2013). "Africa's Contributions to Information Systems," *The African Journal of Information Systems*: 5(4), Article 1.

Watson, R. T., Pitt, L. F., and Kavan, C. B. (1998). Measuring IS Quality: Lessons from Two Longitudinal Case Studies. *MIS Quarterly*, 61-79.

Whitson, T.L. and Davis, L. (2001). Best Practices in Electronic Government: Comprehensive Electronic Information Dissemination for Science and Technology, *Government Information Quarterly,* 18(2), 79–91.

Yin, R. (1994). Case Study Research: Design and methods (2nd ed.). Beverly Hills, CA: Sage Publishing.

Yin, R. (2003). Case Study Research: Design and methods (3rd ed.). Thousand Oaks, CA: Sage Publishing.

Zou, X., Wang, S., and Fang, D. (2008). A Life-Cycle Risk Management Framework for PPP Infrastructure Projects. *Journal Financial Management of Property and Construction.* 13(2), 123-142.

CHAPTER 6:
FDI-Growth Nexus in Ethiopia: Is there any Causality?

Yesuf M. Awel and Tsehaye Weldegiorgis

1.
INTRODUCTION

The effect of foreign direct investment (FDI) on economic growth has been subject of long debate in development literature. Existing theories can be categorized in to two broad opposing views: One, those that claim negative or neutral effect of FDI on economic growth. For instance, predictions based on neoclassical growth models only affect the level of income without changing the long run growth (de Mello, 1997). Similarly, it is argued that FDI may not be beneficial due to possibility of profit repatriation from FDI recipient country to sending country (Seabra and Flach, 2005). Two, those that claim positive effect of FDI on economic growth mainly rooted in endogenous growth theories. They predict positive effect of FDI on growth as far as FDI generates increasing returns to production via externalities (de Mello, 1997, 1999).

It is argued that FDI enhances economic growth by increasing capital stock (composition effect) or facilitating the international technology transfer (knowledge effect). The technology transfer is possibly channelled through imitation, competition, linkage or training (Kinoshita, 2000; Sjolhom, 1999). That is domestic firms may imitate the technology used by multinational companies (MNCs) that operate in the domestic country; they may also be encouraged to use modern technology in order to survive the intense competition in the market due to entrance of MNCs in the domestic economy. Besides, MNCs may transact with domestic firms subcontracting some of their activities or requiring for intermediate inputs from domestic firms creating linkage that lead to different forms of assistance like changing their production process or training the domestic firms' human capital to use the new production process or technology.

Both theoretical views cannot be ruled out without empirical verifications. In line with this, there are several empirical studies examining the relationship between FDI and growth employing variants of econometric methods. However, the evidence is mixed. For instance, Liu et al (2002) reported positive effect of FDI on economic growth in China. Zhang (2001) examines the causality between FDI and growth for eleven countries on country-by-country basis and found mixed evidence. Carkovic and Levine (2002) reported no long run effect running from FDI to economic growth. Recently, Chowdhury and Mavratos (2006) found no causality link between FDI and growth in Chile, while present in Thailand and Malysia. Similarly, Hansen and Rand (2006) found no long run relationship between FDI and economic growth. However, they found causality running from FDI to gross capital formation (GCF) ratio to GDP. They used the latter indicator of FDI to separately test whether the knowledge or composition effect operates. Detailed survey of the relationship between FDI and economic growth is beyond the scope of this paper (see de Mello (1997, 1999); Ozturk, 2007).

Methodologically, much of the FDI-Growth literature is dominated by panel data based growth regression in view of capturing both cross-country and time dimensions (de Mello, 1997; Ozturk, 2007). The results from such regression indicate heterogeneous country impacts that imply for investigation at country level (Chowdhury and Mavrotas, 2006). Besides, such regressions have disadvantage of being subjected by too much averaging. Thus, FDI-growth nexus would be better investigated for a specific country using time-series approaches. Following this argument, we examine the causal link between FDI and economic growth in Ethiopia. Moreover, Ethiopia is an interesting case due to recent huge inflow of FDI (FDI rose from almost nil to nearly 5.5% of the Gross domestic Product [GDP] over last two decade), the recent policy attentions towards attracting FDI and the good growth record over the last 7 years. Besides, to the best of our knowledge, there is no previous empirical study that examines the causality between FDI and growth in Ethiopia. In terms of contribution, the paper add to the empirical literature on FDI-growth nexus drawing evidence from SSA country, Ethiopia, and employing the more robust Toda-Yamamoto (1995) causality test following (Chowdhury and Mavrotas, 2006).

2.
METHODOLOGY

2.1 The Data

According to UNCTADSTAT (2011), foreign direct investment (FDI) is defined as "an investment involving a long-term relationship and reflecting a lasting interest in and control by a resident entity in one economy (foreign direct investor or parent enterprise) of an enterprise resident in a different economy (FDI enterprise or affiliate enterprise or foreign affiliate). Such investment involves both the initial transaction between the two entities and all subsequent transactions between them and among foreign affiliates".

We used annual data on FDI and Growth for the period 1974-2010. FDI is measured as foreign direct investment as percentage of GDP. While growth is measured as Real GDP, real GDP per capita or the rate of growth in real GDP per capita. The main source of data is UNCTADSTAT (2011). Summary statistics of the variables used in the analysis of this study are provided in table 2.1 below.

Table 2.1: Data Description of the Main Variables

Variables	Description	Mean	Standard Deviation
RGDP	Real Gross Domestic Product	8645.16	3733.48
RGDPPC	Real Gross Domestic Product per capita	154.45	27.58
FDI	Foreign Domestic Investment to GDP ratio	1.05	1.62

Source: UNCTADSTAT (2011)

2.2. Econometric Approach

2.2.1. Stationarity Test

An issue in the estimation of time series data is to determine whether the variables are stationary or not, hence one required to test whether the two variables are non-stationary or not, i.e., a unit root test. This is due to the fact that simple regression of two independent nonstationary series often result in a significant t statistics that indicates a statistically acceptable relationship while there is no sense in which the two variables to be related, spurious regression case (Granger and Newbold, 1974). We used the Augmented Dickey Fuller (ADF) test to check whether the two variables are nonstationary or not as formulated in (1) below:

$$\Delta y_t = \gamma_0 + \gamma y_{t-1} + \sum_{i=1}^{k} \gamma_i \Delta y_{t-i} + v_t \qquad (1)$$

Where: $y_t = (gdp_t, fdi_t)$ the variables to be tested for non-stationarity. The null hypothesis is the variables GDP and FDI contain Unit root and the alternative is each are stationary [i.e., integrated of order 0, I(0)]. If γ is statistically significant y_t is I(0). Here the ADF test requires use of the Dickey Fuller critical value.

If the series of the variables contain Unit root, one may suspect spurious regression problem between GDP and FDI. However, the time series literature have shown that regressing an I(1) dependent variable on an I(1) independent variable can be informative if these variables are related in some particular sense, that is, if they are co-integrated.

Asking whether two economic time series variables are cointegrated is like asking whether any long run relationship exists between the trends in the two variables. Given an I(1) gdp_t and I(1) is fdi_t, we test for cointegration using Johansen Maximum Likelihood (ML) procedure.

2.2.2. Johansen ML Cointegration Test

We used the Johansen cointegration procedure for conducting cointegration regression analysis since it provides a unified method for estimating and testing cointegrating relations in the framework of vector error correction (VEC) models (for details, see Johansen 1988; Johansen and Juselius 1990; Enders 1995, Harris and Sollis, 2003).

In a multivariate framework, following Harris and Sollis (2003), if one defines a vector $Z_t = (z_{1t}, z_{2t}, ..., z_{tn})$ and allow each variable to be potentially endogenous in the system. Then one can write a VAR model with k lag length as in (2) below. In this study,

$$Z_t = \begin{bmatrix} h\ GDP_t \\ FDI_t \end{bmatrix}$$

$$Z_t = v + A_1 Z_{t-1} + \ldots + A_k Z_{t-k} + U_t \qquad U_t \sim N(0, \Sigma)$$
(2)

Equation (2) can be reformulated in to a vector error correction form as in (3) below:

$$\Delta Z_t = v + \sum_{i=1}^{k-1} \Gamma_i \Delta Z_{t-i} + \Pi Z_{t-1} + U_t$$
(3)

Where $\Pi = \sum_{j=1}^{j=k} A_j - I$, $\Gamma_i = -\sum_{j=i+1}^{j=k} A_j$, v and U_t are vector of parameters and vector of i.i.d normal disturbance terms. Engle and Granger (1987) show that if the variables Z_t are I(1) the matrix Π in (3) has rank $0 \leq r < m$, where r is the number of linearly independent cointegrating vectors and m is the number of endogenous variables in the system. Hence, one can express Π as $\Pi = \alpha\beta'$, where α and β are both $m \times r$ matrices of rank r.

Specifying the system as in (3) contains information on both the short run and long run adjustment to changes in Z_t via the estimates of Γ_i and Π ($\hat{\Gamma}_i$ and $\hat{\Pi}$). Allowing for linear trend, a constant and assuming that there are r cointegrating relationships; one can rewrite (3) as follows:

$$\Delta Z_t = \alpha(\beta Z_{t-1} + \mu + \rho t) + \sum_{i=1}^{k-1} \Gamma_i \Delta Z_{t-i} + \gamma + \delta t + U_t$$
(4)

By putting restrictions on the trend terms in (4), one can have 5 different vector error correction models. Identifying the proper specification among these 5 different specifications is an important issue in the Johansen procedure; since improper specification of the deterministic component (trend and constant terms) results in inference problem (Hendry and Juselius, 2000). Johansen (1992) suggests the need to test the joint hypothesis of both the rank order and the deterministic components, based on the so-called *Pantula principle*. That is to move through from the most restrictive alternative to the least restrictive model and at each stage to compare the trace test statistic to its critical value and only stop the first time the null hypothesis is not rejected. We followed the same to identify the proper specification and test for existence of cointegration.

The existence of cointegration implies that there is a long run, causal relationship between the variables of interest, at least in one direction. The

Granger causality tests can, then, be performed within the VEC framework to determine the direction of the causality. However, due to sensitivity of the stationarity and cointegration test in finite sample rendering incorrect inferences (Toda and Yamamoto, 1995), we used the Toda-Yamamoto causality test to check for causal link between FDI and growth following Chowdhury and Mavartos (2006). This approach address the problem by specifying standard VAR in levels of the variables instead of difference; hence, minimizes the risk associated with testing for order of integration and the distortion of the tests' size as a result of pretesting (Mavartos and Kelly, 2001).

2.2.3 Toda-Yamamoto Causality Test

In this test, one need to estimate an augmented VAR($k+dmax$) model, where k is the optimal lag length in the original VAR system, and $dmax$ is the maximal order of integration of the variables in the VAR system. The Granger no-causality test is based on a modified Wald (*MWald*) test for zero restrictions on the parameters of the original VAR(k) model. The remaining $dmax$ autoregressive parameters are regarded as zeros and ignored in the VAR(k) model. This test has an asymptotic chi-square VAR ($k + dmax$) distribution when the augmented VAR is estimated. Rambaldi and Doran (1996) have shown that the *MWald* tests for testing Granger no-causality experience efficiency improvement when Seemingly Unrelated Regression (SUR) models are used in the estimation. We followed the same.

3.
FDI AND ECONOMIC GROWTH IN ETHIOPIA: A DESCRIPTION

For long FDI flow to Ethiopia was very minimal. Perhaps this was due to instability in the country, the closed economic system preferred by the communist party that led the country during 1974-1991. In this section, we describe the trends for two periods 1974-1991 and 1991-2010 periods. The period selection is mainly dictated by the availability of data and the different economic system path the country followed over this two periods.

As can be seen from figure 3.1 (right panel), the period 1974-1991 witnessed deterioration in economic performance. This was a period of control regime, pro-closed market movement and protracted civil war. During this period, the country embarked on nationalization of productive assets, business environment became hostile to investors and markets highly regulated. To once dismay, the country faced 1984/85 famine. All this contributed to the dismal economic performance.

In view of foreign direct investment, the period recorded an average of 0.097% of the GDP that fell down from nearly 1% of the GDP. The political turmoil, the overvalued exchange rate and discriminatory economic policies that penalized private sector clearly sent a message of bad FDI destination.

The period 1991-2010 was a period of recovery and recently witnessing sustained growth over the last seven years. With the down fall of the pro-socialist regime and formation of federal states, the incumbent adopted structural adjustment policies with Ethiopian context of liberalization (Geda and Befekadu, 2005). The military expenditure declined sharply. Black market premium fell from 358% in 1992 to 15.5% in 1997 (Geda and Befekadu, 2005). Overall, the policy reforms and political stability contributed towards better economic performance. Notwithstanding the down turns during 1994 drought, 1998 Ethio-Eritrean border conflict and other political instabilities.

Looking in to FDI, the period attracted huge flow of FDI that once reached to nearly 5.5% of the GDP. This was possible due to the liberalization of different markets that attracted foreign investors, devaluation of Birr (Ethiopian currency) and different investment proclamations with preferential treatments (like tax holidays) to attract investment. Despite the average rise in foreign investment, the FDI flow was highly volatile as can be seen in figure 3.1 below. Partly, this was due to frequent revisions of investment proclamation and instability due to border conflict with Eritrea that perhaps reduced investment confidence.

Looking in to the trend of FDI and growth pre 1991(figure 3.1), it seems that one cannot establish a link between FDI and growth. However, visualizing the relationship post 1991a positive link could be shown on average though a negative relationship emerge in the last couple of years.

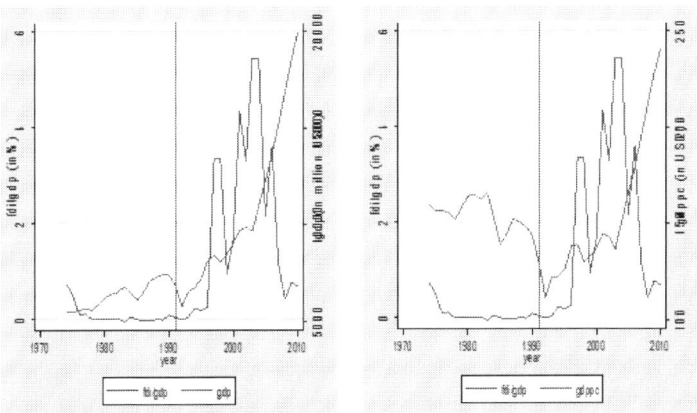

Figure 3.1: Trend of FDI-to-GDP ratio (fdi/gdp) , Real GDP (gdp) and Real GDP percapita (gdppc) [1974-2010]

4.
ECONOMETRIC RESULTS

The econometric results are given in the following four steps. First, we present the ADF unit root test. Second, we present the optimum lag structure selection for the VAR. Third, the Toda-Yamamoto causality test results are provided. Last, cointegration test result based on Johanson ML procedure also given.

Table 4.1 provides ADF test results. The results show that both real GDP and FDI are I(1) series at 5% level of significance. The test for two unit roots, I(2) is rejected at 10% level of significance, hence the two variables are integrated of order one. As can be seen from table 4.1, disaggregating the data in to two different regimes, the results hold true.

Table 4.1: ADF Test Results♦

Variables	Whole Period: 1974-2010		Period I: 1974-1991		Period II: 1991-2010	
	Ho: I(1)	Ho: I(2)	Ho: I(1)	Ho: I(2)	Ho: I(1)	Ho: I(2)
lnRGDP	-0.297	-4.338***	-3.315	-5.780***	-0.435	-3.149*
	(0.9896)	(0.0027)	(0.0639)	(0.0000)	(0.9858)	(0.0950)
FDI to GDP Ratio	-2.191	-5.763***	-10.967	-3.078	-1.389	-4.393***
	(0.4951)	(0.0000)	(0.0000)	(0.1115)	(0.8640)	(0.0022)

Values in parenthesis are MacKinnon approximate p-value

♦The lag length for the ADF test was set at one lag length based on Schwarz Bayseian Information Criteria (SBIC)

The results for selecting the optimum lag structure of the VAR (k) are found based on the Akaike information criteria (AIC), Hannan-Quinn information criteria (HQIC) and SBIC. All consistently indicate the optimum lag length of 1. Hence, we selected one lag as the optimum lag length for the VAR model to be estimated.

Given the maximal order of integration (d_{max}=1) and the selected optimal VAR length (k=1), we estimated the following augmented VAR model using Seemingly Unrelated Regression (SUR) method.

$$Z_t = v + A_1 Z_{t-1} + A_2 Z_{t-2} + U_t \qquad U_t \sim N(0, \Sigma)$$
(5)

Where: $Z_t = \begin{bmatrix} h\ GDP_t \\ FDI_t \end{bmatrix}$, $Z_{t-1} = \begin{bmatrix} h\ GDP_{t-1} \\ FDI_{t-1} \end{bmatrix}$ and $Z_{t-2} = \begin{bmatrix} h\ GDP_{t-2} \\ FDI_{t-2} \end{bmatrix}$.

$A_1 = \begin{bmatrix} \alpha_1^1 & \alpha_2^1 \\ \alpha_2^1 & \alpha_1^1 \end{bmatrix}$ and $A_2 = \begin{bmatrix} \alpha_1^2 & \alpha_2^2 \\ \alpha_2^2 & \alpha_1^2 \end{bmatrix}$.

Testing for causality based on Toda-Yamamoto (1995) is to check whether $\alpha_2^1 = 0$ in the case of FDI does not granger cause growth and $\alpha_2^1 = 0$ in the case of growth does not granger cause FDI. Table 4.2 below presents the result.

Table 4.2: Toda-Yamamoto Causality Test Result

Null Hypothesis	1974-2010	1974-1991	1991-2010
	MWald Test	MWald Test	MWald Test
FDI doesnot granger cause GDP	0.02 (0.8744)	0.81 (0.3676)	1.45 (0.2282)
GDPdoesnot granger cause FDI	0.21 (0.6446)	1.69 (0.1931)	0.05 (0.8217)

Values in parenthesis are p-value

The evidence clearly portray for not rejecting the null hypothesis of Granger non-causality from FDI to growth and vice versa. We could not find a causal link between FDI and Growth in Ethiopia. Finding the Granger non-causality between FDI and Growth and the I(1) series of the two variables, we further checked for any long run relationship between FDI and growth. The results for Johanson ML cointegration test is provided in table 4.3 below.

Table 4.3: Johanson ML Cointegration Test result for FDI and Growth
[Restricted constant Specification]

Null Hypothesis	Maximum Eigen Value	5% critical value	Trace Statistics	5% critical value
1974-2010				
r=0	15.5140	15.67	20.5357	19.96

r≤1	5.0218	9.24	5.0218**	9.42
1974-1991				
r=0	23.9382	15.67	29.7860	19.96
r≤1	5.8479	9.24	5.8479**	9.42
1991-2010				
r=0	10.7433	15.67		19.96
r≤1	4.3893	9.24	4.3893	9.42

The cointegration test reveals that possibly there is one cointegration rank implying a long run relationship between FDI and growth in Ethiopia. Variant of specifications of vector error correction model with different indicators of growth variable tested and the cointegration rank is the same in almost all cases. This result may imply two phenomenon. One, it may be the case that the variable FDI is a very good proxy for a variable that affect growth. Two, it may be the case that the aggregate level of FDI is too small to translate in to growth in our sample period but a possible indication of the relationship over the long run. Further in ensuring the robustness of the results presented hitherto. We used different indicators of the variables FDI (FDI to GCF ratio) and Growth (real GDPPC, RGDPPCgr). Though in the interest of brevity not presented here, the results are qualitatively the same and could be obtained from the author up on request. Further, one would like to examine the growth performance in a multivariate framework incorporating other possible endogenous variables. The availability of data was constraining hence lest for small sample size that could render fragile statistical results in multivariate time series framework. Moreover, this study does not explicitly address issue of structural breaks that may limit the results, hence a caution in drawing conclusions. With further availability of data, these can be possible future research area.

5.
CONCLUSION

The objective of this paper was to investigate the causal link between FDI and economic growth in Ethiopia. Using annual data ranging from 1974-2010 and employing the Toda-Yamamoto (1995) bivariate causality test, we could not find any causality running from FDI to growth or vice versa. However, there was an evidence of cointegration between FDI and growth. That is, possible long run relationship between FDI and growth. The implications of the results are: first, the flow of the aggregate level of FDI is too small to translate in to growth. This may indicate that economic growth is a necessary but not sufficient condition for attracting FDI. Thus, efforts should be exerted on addressing the fundamentals (like infrastructure, human capital and institutions, etc.) than specific policy for attracting FDI. Second, perhaps FDI flow has gone in to sectors that could not create linkages and fuel economic growth.

This may indicate the impact of FDI on growth is conditional on the type of economic sector, where technology transfer, managerial skills, organizational process and skill acquisitions are more possible than otherwise. Thus, developing countries like Ethiopia should formulate policies attracting FDI in to economic sectors that could harness the benefits of the FDI outweighing the costs of hosting FDI (like profit repatriation to FDI sending economy). Seabra and Flach (2005) document adverse impact of profit repatriation due to FDI. As pointed out earlier, with further availability of data future research should examine the causal link in a multivariate framework while addressing the issue of structural break to uncover more robust evidence.

REFERENCES

Carkovic, M. and Levine, R., (2002), 'Does Foreign Direct Investment Accelerate Economic Growth?', Working Paper Series. http://papers.ssrn.com/sol3/papers.cfm?abstract_id=314924. [Accessed October 29, 2011]

Chowdhury, A. and Mavrotas, G., (2006), 'FDI and Growth: What Causes what?', *The World Economy*, 29 (1), 9-19.

de Mello, L. R., (1997). 'Foreign Direct Investment in Developing Countries and Growth: A Selective Survey', *Journal of Development Studies*, 34(1), 1–34.

de Mello, L. R., (1999). 'Foreign Direct Investment-led Growth: Evidence from Time Series and Panel Data', *Oxford Economic Papers*, 51(1), 133–51.

Enders, W., (1995). *Applied Econometric Time Series,* New York, NY: John Wiley and Sons.

Engle, R. F. and Granger, C. W. J. (1987). 'Cointegration and Error Correction: Representation, Estimation, and Testing', *Econometrica,* vol. 55(2), 251-76.

Geda, Alemayehu and Degefe, B., (2003). 'Explaining African Economic Growth: The Case of Ethiopia', Final Report to AERC. http://alemayehu.com/AA%20Recent%20Publication/Growth_FinalFR_August_2005.pdf [Accessed October 25, 2011].

Granger, C. W. J., and Newbold, P., (1974). 'Spurious Regressions in Econometrics', *Journal of Econometrics,* 2, 111–120.

Hansen, H. and Rand, J., (2006). 'FDI and Growth: What Causes what?', *The World Economy,* 29(1), 21-41.

Harris, R. and R. Sollis (2003), Applied Time Series Modelling and Forecasting, Chichester, John Wiley and Sons.

Hendry, D. and Juselius, K. (2000), 'Explaining Cointegration Analysis: Part II.' Discussion Papers 00-20, Department of Economics, University of Copenhagen, Denmark.

Johansen, S. (1992), 'Determination of cointegration rank in the presence of a linear trend', *Oxford Bulletin of Economics and Statistics,* 54(3), 383-397.

Johansen, S. (1988), 'Statistical analysis of Co-integrating Vectors', *Journal of Economic Dynamics and Control,* 12(2-3), 231-254.

Johansen, S. and K. Juselius (1990), 'Maximum Likelihood Estimation and Inference on Cointegration-with Applications to Demand for Money', *Oxford Bulletin of Economics and Statistics,* 52(2), 169-210.

Kinoshita, Y. (2000), 'R&D and technology spill over via FDI: Innovation and absorptive capacity', Accessed from http://papers.ssrn.com/sol3/papers.cfm?abstract_id=258194 [Accessed 29 October 2011]

Liu, X., Burridge, P. and Sinclair, P.J.N., (2002), 'Relationships between economic growth, foreign direct investment and trade: Evidence from China', *Applied Economics,* 34(11), 1433-1440.

Mavrotas, G. and R. Kelly (2001), 'Old Wine in New Bottles: Testing Causality Between Savings and Growth', *The Manchester School,* 69(1), 97–105.

Ozturk, Ilhan (2007), 'Foreign Direct Investment – Growth Nexus: A Review of the Recent Literature', *International Journal of Applied Econometrics and Quantitative Studies* ,4(2), 79-98. http://www.usc.es/~economet/reviews/ijaeqs424.pdf [Accessed 29/10/2011].

Rambaldi, A.N., and Doran, H.E. (1996), 'Testing for Granger non-causality in cointegrated systems made easy', Working Papers in Econometrics and Applied Statistics 88, Department of Econometrics, The University of New England.

Seabra, F and L. Flach (2005), 'Foreign Direct investment and Profit Outflows: A Causality Analysis for the Brazilian Economy', *Economics Bulletin,* 6(1), 1-15.

Sjoholm, F., (1999), 'Technology gap, competition, and spill overs from FDI: evidence from establishment data', *Journal of Development Studies*, 36(1), 53–73.

Toda, H. Y. and T. Yamamoto (1995), 'Statistical Inference in Vector Auto regressions with Possible Integrated Processes', *Journal of Econometrics*, 66(1), 225–50.

Zhang, K. H. (2001), 'Does Foreign Direct Investment Promote Economic Growth? Evidence from East Asia and Latin America', *Contemporary Economic Policy*, 19(2), 175–85.

CHAPTER 7:
Scenario Planning as a Management Tool for Sustainable Aquaculture

Worku Jimma, Daniel Adjei-Boateng and

Nicolas Van Vosselen

1.
INTRODUCTION

Aquaculture is the fastest-growing sector of the world food economy, with 10-12 percent rate of increase each year, compared with only 1.4% for capture fisheries and 2.8% for terrestrial meat production systems (FAO, 2003). As the world's human population continues to expand beyond 6 billion, its reliance on farmed fish as an important source of protein continues to increase (Naylor *et al.*, 2000). It is expected that by 2030, aquaculture will dominate fish supplies and more than half of the fish consumed is likely to originate from this sector (FAO, 2000). Enhancing food security and alleviating poverty are major and complementary global priorities. Aquaculture has a special role in achieving these objectives because, firstly, fish is a highly nutritious food that forms an essential, if not indispensable, part of the diet of a large proportion of people in developing countries. Secondly, it could improve economic activity of rural communities by converting marginal lands into ponds for fish farming.

Over the years various aspects of this industry have come under scrutiny due to their impacts on communities, the aquatic environment, other aquatic organisms, diseases, pollution and more importantly the sustainability of the aquaculture industry itself with regards to its impacts on the environment and utilisation of resources. An important issue that has always come up for discussion is the reliance on fishmeal in fish feeds. Fishmeal is produced by industrially converting small-size fish into meal. These small-size fish which could be used for human consumption is some regions of the world is converted to fishmeal due the economics of scale. Due to the high cost of fishmeal, fish feeds accounts for 50-60% of aquaculture production cost. Coupled with this high cost of production are the issues of sustainability, the environmental impacts and the ethical issue of catching fish to feed fish.

In order to ensure the survival of the aquaculture industry, it is imperative that the principles of sustainable aquaculture as defined by the FAO, the European Commission, and environmental bodies such as World Wide Fund for nature are adhered to. Sustainable aquaculture takes its roots from sustainable development as defined by the Brundtland commission (1987), as development in which present generations find ways to satisfy their needs without compromising the chances of future generations to satisfy theirs. Few would disagree with this definition. While we all agree that sustainability is essential, the difficulty arises in defining the term in relation to aquaculture. Two aspects of the aquaculture industry that tend to preoccupy its critics are the environmental impacts, and the use of 'industrial fisheries' to provide fishmeal and fish oil to feed piscivorous fish. Researchers involved with aquaculture however, points out that, there is the need to look at the whole process, including the market, productivity, social equity, economic feasibility as well as the physical impacts of aquaculture. This is further complicated by the sheer diversity of the sector, which ranges from small backyard fishponds to large, highly commercialised and automated operations.

Scenario planning is a systematic method for thinking creatively about possible complex and uncertain futures. The central idea of scenario planning is to consider a variety of possible futures that include many of the important uncertainties in the system rather to focus the accurate prediction of a single a single outcome (Peterson et al. 2003). A scenario is defined as a structured account of a possible future. Scenarios describe futures that could be rather than futures that will be (Van der Heijden, 1996). In essence, scenarios are alternative dynamic stories that capture key ingredients of our uncertainty about the future of a study system. Scenarios are constructed to provide insight into drivers of change, reveal the implications of current trajectories and illuminate options for action. Scenario planning is somewhat similar to adaptive management (Walters, 1986), an approach to management that takes uncertainty into account. Scenario planning is most useful when there is a high level of uncertainty about the system of interest and system manipulations are difficult. Therefore this study was conducted with the aim to create scenarios on the internal driving forces that could shape the aquaculture practices and direct it on the path of sustainability.

2.
METHODOLOGY

2.1 Scenario Planning

Scenarios offer an alternative environment in which today's decisions may be played out. They are neither predictions nor strategies; they are descriptions of possible futures with an emphasis on events and trends. Scenarios are designed to highlight opportunities and risks inherent in specific strategic issues.

Three main reasons exist for choosing the scenario methodology as the vehicle for this study:

- **Anticipation and leverage of change**: The scenarios will help identify surprises and discontinuities in trends and those elements identified as crucial to supporting the reform/ transition process. As such, it should be easier to identify pitfalls and provide possibilities to leverage new opportunities as well as produce robust and resilient strategies for the future;

- **Stimulate new ways of thinking**: Scenarios encourage thinking beyond traditional approaches to problem solving and exploitation of opportunities. They have the power to break stereotypes and this new way of thinking can serve as a catalyst for radical changes.

- **Reducing future risk:** The use of scenarios can help the aquaculture industry better determine the outcomes of certain actions before they are actually taken.

3.
SCENARIOS FOR SUSTAINABLE AQUACULTURE

3.1 Scenario Planning on Aquaculture in 2050

Aquaculture is the fastest-growing sector of the world food economy, with 10-12 % rate of increase each year. As the world's human population continues to expand beyond 7 billion, reliance on farmed fish as source of protein continues to increase. It is expected that by 2030, aquaculture will dominate world fish supplies and more than half of the fish consumed is likely to originate from this sector (FAO, 2000). This prediction is based on the projections on current production levels as shown in figure 1.

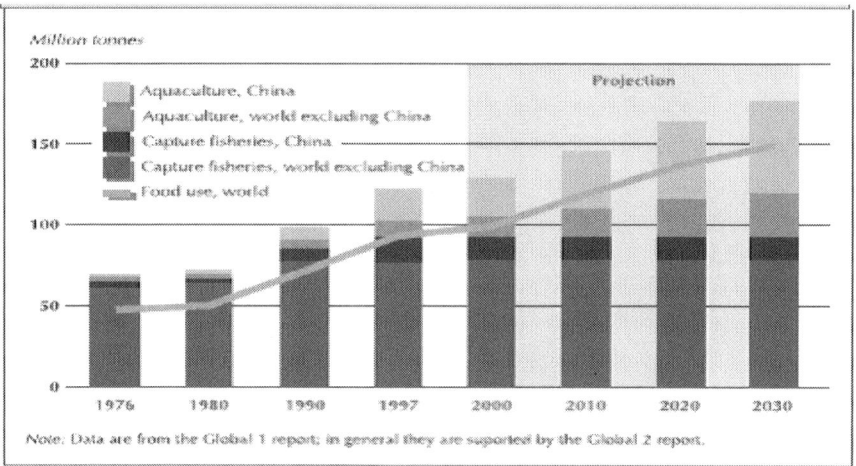

Figure 1. World Fish Production and Food use consumption 1976-2030.
(Source: *The state of World Fisheries and Aquaculture, FAO report 2002*)

The development of scenarios on the future of aquaculture, an industrial process of raising fish, crustaceans, molluscs and aquatic plants for human consumption and other purposes in the aquatic medium is essential. Several driving forces which were recognised as internal forces (biotechnological innovations in feeds, disease control, reproduction and environmental control) are likely to shape the way aquaculture is practiced in the future. The internal driving forces of aquaculture are changing by the day with new scientific developments. These forces are exerting a strong influence on the physical structures and the manner in which aquaculture is practice today and would be practise in the future.

3.1.1 Scenarios on Internal Driving Forces of Aquaculture

Biotechnological innovations - Disease and environment
Description of Scenario Parameters

The four scenarios below depict the internal driving forces that will characterize aquaculture by 2050 (Figure 2). The development of biotechnological innovations to the fishmeal problem is envisaged where cheap alternatives such as plant proteins, genetically modified plants enriched with the deficient amino acids and single cell protein produced from microorganisms using agro-waste and waste products.

Firstly, biotechnological innovations are expected to lead to a breakthrough in the artificial reproduction of most of preferred fish species such as the cod, tuna, eel and aquarium species whose reproduction in captivity has been a bottleneck till date. Secondly, in the area of disease and environmental control, it is envisaged that disease control will see a lot of advancement. Vaccine development to combat and fight diseases in aquaculture will rapidly accelerate within this period. Similarly, the use of immune-stimulants and probiotics to prop immune response of culture fish will become a routine fish husbandry practice, as the emphasis will shift from disease control to prevention.

Environmental control will foresee the evolution of culture systems (tank systems, cages, recirculation systems, ponds) that effectively treat their wastes and effluents thus, completely eliminating the environmental pollution that are characteristic of present systems. Sustainable aquaculture will be the order of the day as aquaculture will be a net producer of fish and not just a through-put of converting one type of fish in the form of fishmeal to a preferred fish.

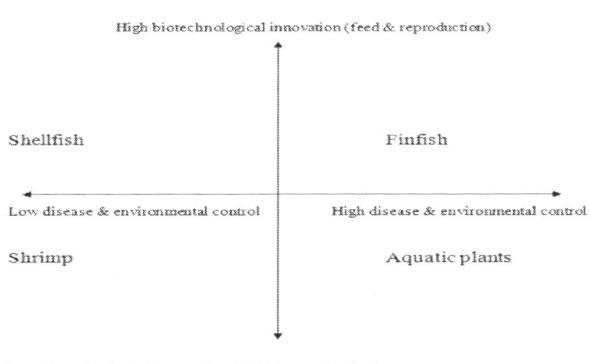

Figure 2. Scenario Matrix

Shrimp Scenario

Low biotechnological innovation, low disease and environmental control.

This shrimp scenario explores the possible outcome of a situation in which aquaculture is practiced currently within the confines of present knowledge and technology. Fishmeal continues to be the primary protein source in fish feed, the reproduction of preferred species such as cod, tuna, eel continues to be a bottleneck.

In the field of disease control, not much has been achieved, disease outbreaks are rampant and antibiotics have failed to protect culture fish from disease outbreaks due to antibiotic resistance. Environmentally, aquaculture continues to pollution aquatic environment, the wild harvest of fingerlings and the destruction of mangroves to establish shrimp ponds continues unabated with serious effects on the recruitment of natural fish stocks, erosion of coastal areas and acidification problem.

The shrimp scenario is therefore one of doom, where aquaculture activities, which fails to take proper care of the environment in which it operates as well as evolve sustainable technologies will eventually become a backlash on its viability and sustainability.

Aquatic plants Scenario

Low biotechnological innovation, high disease and environmental control.

The aquatic plants scenario portrays a situation where not much has been achieved biotechnologically to de-emphasize the dependence of aquaculture on fishmeal. As a result, fishmeal continues to be the primary protein source in fish feed. The reproduction of preferred species such as cod, tuna, eel continue to be a bottleneck.

Advances in vaccine technology in combination with other preventive disease control measures such as the use of probiotics to suppress the proliferation of harmful bacteria and viruses have enable a greater control of disease outbreaks. The use of immune-stimulants to prop the immune system of culture fish against noxious microorganisms has to prevent disease outbreaks or kept their occurrence to the minimum. Strict adherence to environmental regulations governing cage aquaculture operations have succeeded in minimizing the polluting effects of cage aquaculture.

However, fishmeal continues to be used and the wild capture of fingerlings of species whose reproduction continues to pose a problem is ongoing, the environmental impacts of aquaculture on wild fish populations have reached a disturbing stage.

The aquatic plants scenario is therefore an unhealthy development since advances made in the field is one-sided. It can be likened to the current state of development of salmon aquaculture, *i.e.,* high reproductive success but continues to pollute the environment due to its continuous use of fishmeal and its escapee problems.

Shellfish Scenario

High biotechnological innovation, Low disease and environmental control

The shellfish scenario envisages the situation where there is high biotechnological advancement that has unravelled the feed and reproduction bottleneck. Fishmeal is no longer the primary source of protein in fish feeds. It has been totally replaced with alternatives such as plant proteins and single cell proteins from microorganisms cultured cheaply with agro-processing wastes. The artificial reproduction of hitherto difficult to reproduce species such as the eel, cod and coral reef species has been successfully achieved. The pressure on wild fish stocks does not exist anymore. No seed collection from the wild and no threat to biodiversity.

However, earlier reliance on antibiotics and other chemical agents for disease control has resulted in the emergence of resistant and virulent viruses and bacteria. The consequence of this is frequent outbreaks of disease with no remedy to curb their spread. The low water quality also facilitates the spread of diseases.

In view of the high disease levels and low water quality, only hardy and disease resistant species such as catfish could be successfully farmed in the polluted waters. Species that require high water quality for survival have all been eliminated.

The shellfish scenario is also a reflection of a lopsided development that will in the long-term result in the creation of a culture environment conducive to the growth of species that feed low on food chain.

Finfish Scenario

High biotechnological innovation, high disease and environmental control

The finfish scenario explores the situation where both biotechnological innovations and disease & environmental control are highly advanced and are moving at the same pace.

Aquaculture is riding high, all the cultured species can be reproduced artificially and many new species have been domesticated. All life cycle of all aquaculture species can be completed in captivity. Fishmeal has been totally and completely replaces from fish feed formulations. All protein sources are either from plant and microorganisms or are synthesized chemically.

This has enabled the farming of highly preferred and piscivorous species such as the tuna. Tuna can now be reproduced artificially and grow out is no longer a problem since a cheap source protein is available to feed this highly carnivorous species.

Environmentally, recirculation systems has advanced to such an extent that the system treats all of its soluble wastes. The solid wastes are converted to usable microorganism in bio-rectors incorporated in the system. The microorganisms are channelled back into the system as feed for the cultured fish such that nothing is wasted.

Cage culture in bays and near shores have been revolutionized such that the system treats its waste by collecting uneaten feed and faeces from the bottom of the cage and passes it through a treatment bio-rector included in the cage design. As a result of these advances disease outbreaks are absent. Disease control is further enhanced by the use of immune-stimulants, probiotics and vaccinations.

The overall effect is that aquaculture is a net producer of high quality fish for human consumption as well as operates and creates a cleaner environment for other aquatic species to thrive and survive.

4.
CONCLUSIONS AND RECOMMENDATIONS

The past two decades have seen the growth of the aquaculture industry from an experimental/pilot stage to a fully grown important sub-sector. Long-term growth of the aquaculture industry depends on both ecologically sound practices and sustainable resource management. Governments can encourage such practices by stringently regulating the creation of new farming facilities in mangroves and other coastal wetlands, establishing fines to minimise escapes of fish from aquaculture pens, enforcing strict disease control measures for the movement of stock, and mandating effluent treatment and in-pond recirculation of wastewater. Despite significant improvements in the industry, many ecologically sound technologies remain on shelf. This is an arena where external-funding agencies such as development banks can play a strategic role by encouraging the development and financing the implementation of sustainable aquaculture technologies, the rehabilitation of ecosystems degraded by aquaculture, and the protection of coastal ecosystems. We therefore recommend that governmental and development organizations, as well as the aquaculture industry and all its stakeholders promote farming of herbivorous and omnivorous species; reduction and replacement of fish meal and fish oil inputs in feed; development of integrated farming systems that use multiple species to reduce costs and wastes and increase productivity.

REFERENCES

Brundtland Commission, (1987). Our Common Future. New York, Oxford University Press.

FAO, 2000. The State of the World Fisheries and Aquaculture, 1999. FAO, Rome. pp 142.

FAO, 2002. Use of fishmeal and fishoil in Aquafeeds: Further thoughts on the fishmeal trap, by M.B. New and U.N. Wijkstrom. FAO Fisheries Circular No. 975/FIPP/C975.

FAO, 2003. The State of the World Fisheries and Aquaculture, 2002. FAO, Rome. pp 150.

Naylor, R.L., Goldberg, R.J., Primavera, J.H., Kautsky, N., Beveridge, M.C., Clay, J.,

Folk, C., Lubchenco, J., Mooney, H. and Troell, M., (2000). Effect of aquaculture on

world fish supplies. *Nature*, 405, 1017–1024.

New, M., (1999). National Aquaculture Policies, with special reference to Namibia in pp. 303 - 318 Legislation for Sustainable Commercial Aquaculture, Balkema, Rotterdam

Peterson D., Cumming S. and Carpenter R., (2003). Scenario Planning: a Tool for Conservation in an Uncertain World. *Conservation Biology*, 17(2), 358-366.

Van der Heijden, K., (1996). Scenarios: the art of strategic conversation. New York, Wiley.

Walters, C. J., (1986). Adaptive management of renewable resources. New York, Macmillan.

CHAPTER 8:

A Study of a Public-Private Acquisition System applying Transaction Cost Economics Principles

Ermias Kebede

1.
INTRODUCTION

To understand the nature of transaction costs in bilateral exchange relationships the study focused on the specific example of the way MOD engages with Industry. A thorough literature review of defence industrial policy was undertaken. This review identified a clear connection with the procurement policies of the past and the effect that it has had on defence transactions today. The monopolistic nature of the defence market originates from the privatisation policy of the Conservative Government, led by Margaret Thatcher. The competitive adversarial approach to Industry by the MOD has been attributed as a consequence of Peter Levene's (then Chief of Defence Procurement) reforms in the 1980s and further reforms in the 1990s (Levene, 1987; McIntosh, 1993; Bishop, 1995; Humphries and Wilding, 2001). The election of the New Labour government in 1997 and its subsequent defence procurement reforms, through the Strategic Defence Review in 1998, introduced the concepts of Smart Procurement and Integrated Project Teams, attempting to normalise the relationship with industry and having a major impact on the defence industrial strategy.

While there is currently a sea change in defence policy for the purpose of this chapter the focus will remain that of the original doctoral study and will be mainly focused on reforms up to the Strategic Defence and Security Review (SDSR) in 2010. This is also due to the authors view that the SDSR and the subsequent Lord Leven Report (MOD, 2011) places minimal emphasis on acquisition policy in terms of industrial strategy, rather it is based more on MOD structure and management (as stated in the report title). Thus, for our purpose the transactional relationship characteristics of defence acquisition can still be understood applying the same perspective as that of the original doctoral research and consequent thesis of 2011.

The results of the transformations in defence procurement policy of the twentieth century have been to create an industrial engagement, which has been characterised by an adversarial relationship. This has led to the challenges in major defence projects, which are that they, on average, have failed to meet the cost, time and user requirement targets set out in the project planning phase. The proposed explanation for the difficulties in defence projects is directed towards the transactional conditions of major defence projects. In order to understand these transactional consideration and achieve the objectives of the research Transaction Cost Economics (TCE) theory has been applied to the UK defence acquisition system. However, this can be easily deployed to other sectors with similar monopsony-monopoly characteristics, and TCE itself is apt for understanding any industrial exchange conditions.

The objectives of the research were:

- To understand the common factors of *why* some defence projects fail to satisfy value for money criteria such as time, cost and user requirements;
- To highlight common characteristics of defence projects in order to understand *how* defence acquisition can be improved;
- To apply a transaction cost economics lens to the content analysis findings in order to gain *insight* into defence acquisition.

2.
METHODOLOGY

The research was undertaken through a content analysis methodology with the aid of qualitative research software NVivo 8. In the first instance, Major Project Reports during the reporting periods of 1993 to 2010 were systematically analysed to understand the performance of major acquisition projects, and key factors within the defence acquisition system. The second stage of the research analysis moved towards a focus on individual projects and, thus the researcher sought access to Value for Money reports, which provide a more detailed evaluation of the acquisition process with a focus on the value gained for the taxpayers. The more recent reports were accessed through the NAO website (as were all Major Project Reports); however the older reports were provided by the NAO through their internal database.

The Value for Money reports were uploaded to the NVivo 8 software in order to analyse the data using a content analyses approach of categorisation, factorisation and inferences. The findings of the analysis are provided partly through a quantitative presentation of frequency counts, however the findings were mainly used for a qualitative analysis using the application of TCE. The dataset used in the doctoral thesis were Value for Money reports focusing on defence acquisition projects. These reports provide a clear analysis of the process behind the acquisition of major defence equipment, and as such evaluate the way the MOD and its Prime Contractors interact in the tender/bidding, contracting and delivery stages. Table 1 provides a list of the reports analysed; the reports date back as early as 1984 and as relatively recent as *2009.*

Table 1: Defence Value for Money Reports

Session	Report	Report Title
Feb 1984	HC 287	Trident Project
Mar 1985	HC 291	The Torpedo Programme
Jul 1993	HC 864	The Awarding of the Contract for the Landing Platform for Helicopters
1994-1995	HC 692	Procurement Lessons for the Common New Generation Frigate
1994-1995	HC 724	Eurofighter 2000
1998-1999	HC 738	The Procurement of Non-Combat Vehicles for the Royal Air Force
1999-2000	HC 328	The Private Finance Initiative: The Contract for the Defence Fixed Telecommunications System
2001-2002	HC 840	Helicopter Logistics
2001-2002	HC 1246	Building an Air Manoeuvre Capability: The Introduction of the Apache Helicopter

2002-2003	HC 90	The Construction of Nuclear Submarine Facilities at Devonport
2005-2006	HC 1050	Delivering Digital Tactical Communications through the Bowman CIP Programme
2006-2007	HC 825	Transforming Logistics Support for Fast Jets
2007-2008	HC 512	Chinook Mk3 Helicopters
2007-2008	HC 627	Hercules C-130 Tactical Fixed Wing Airlift Capability
2007-2008	HC 788	The Defence Information Infrastructure
2007-2008	HC 1115	The United Kingdom's Future Nuclear Deterrent Capability
2008-2009	HC 295	Providing Anti-Air Warfare Capability: the Type 45 Destroyer

The dataset was analysed using a structured systematic process of content analysis and applying qualitative research software NVivo 8. The process began with reading the Value for Money reports to identify the key themes in the reports. Once the key themes were identified a hierarchical tree (shown in Figure 1) was created to categorise the factors characterising the *defence acquisition process*. Three major categories were highlighted as part of the defence acquisition process: Selection Process, Contracting Process and, Uncertainty, Complexity and Risk. Of these categories, further sub-categories were identified in an exhaustive process of data mining. Where the data in the reports did not equate with the defence acquisition process, a separate tree labelled 'Other' was designated for their categorisation in order to ensure a full analysis was undertaken.

Figure 1: Defence Acquisition Tree Model

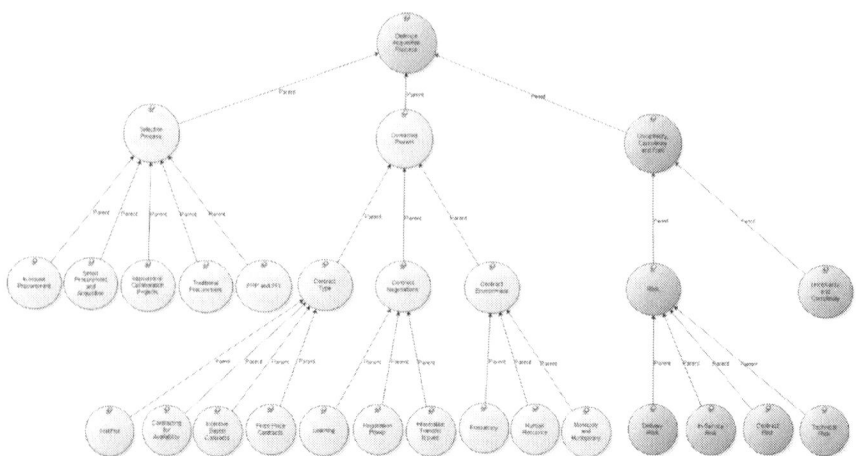

3.
TRANSACTION COST ECONOMICS

The research relied upon TCE theory as a basis of understanding the relationship between the buyer (MOD) and supplier(s) (Industry). TCE is an economic theory developed by Oliver Williamson emanating from his 1975 book *Markets and Hierarchies*. The theory develops the conditions under which a buyer and supplier exchange goods/services. The theory suggests that there are environmental factors affecting any transaction, these are Uncertainty/Complexity, Asset Specificity and Frequency, when any of these are present in a transaction they can create a number of consequences. These consequences are manifested through the behavioural assumptions of decision makers in a transaction – these are bounded rationality and opportunism. This research has also relied upon learning as a behavioural assumption, formalised in the TCE model by Winch (1989). The combination of these environmental and human behavioural factors will create a number of transaction governance arrangements in a spectrum from market governance to unilateral governance (internal organisation hierarchies). In this research, the use of bilateral governance (partnering/strategically linked organisations) using a relational contracting approach is suggested for the UK defence acquisition case. Bilateral governance lies closer to the unilateral governance side of the spectrum.

Asset Specificity is identified by Williamson (1985) as the most important of the environmental factors affecting a transaction; it most distinguishes TCE from other economic theories. Williamson identifies four types of asset specificity, with an additional fifth provided by Masten et al. (1991):

1. Site specificity is the geographical distribution of the transaction e.g. the importance of the location of a dockyard.

2. Physical asset specificity is the investment in physical components for a transaction e.g. a target acquisition system on a fighter aircraft.

3. Human asset specificity is the firm-specific knowledge possessed by human resources e.g. systems engineers specialising in networking capability.

4. Dedicated assets are discrete investments in generalised production capacity (as contracted with special purpose) that would not otherwise be made but for the prospect of selling a significant amount of product to a specific customer e.g. a shipyard purposely built to make aircraft carriers.

5. Temporal specificity is the importance of delivering the transaction to a specific deadline e.g. the delivery of urgent operational requirements to the theatre.

Asset specificity plays a large role in the nature of defence transactions. It has been explained in the development of the thesis that the transaction specific investments in the last five decades of defence industrial policy have fostered a monopolistic defence industrial base in the UK.

4.
PROCUREMENT ROUTE

The Selection Process (shown in Figure 1, Branch 1) is the category that captures the procurement route adopted in the first phase of the acquisition of an equipment. This is identified in five finite sub-categories – In-house Procurement, Traditional Procurement (i.e. market-based, arms-length procurement), Public Private Partnership and Private Finance Initiatives (PPP and PFI), and Smart Procurement and Acquisition. These sub-categories can be generalised in most acquisition systems, with the exception of SMART procurement and SMART acquisition approaches, which applies principles specifically developed by the MOD for defence procurement.

The Value for Money reports were analysed with respect to the selection process, passages from the reports which referenced a procurement process was taken and coded into these five categories in NVivo 8. The results presented in the pie chart (Figure 2) illustrate the procurement routes taken by the MOD in the 25 years period of the Value for Money reports. It should be noted that within a single project more than one selection process is possible, since the MOD may have re-competed the project, used a separate procurement route for the support phase or affiliated projects, otherwise it was due to a reassessment of the selection process e.g. many of the traditional procurement's were assigned reapproved phases using the SMART Acquisition Concept, Assessment, Demonstration, Manufacture, In-Service and Disposal (CADMID) project life cycle. These conditions were taken into account when making the classifications for the references to the selection process of an MOD acquisition.

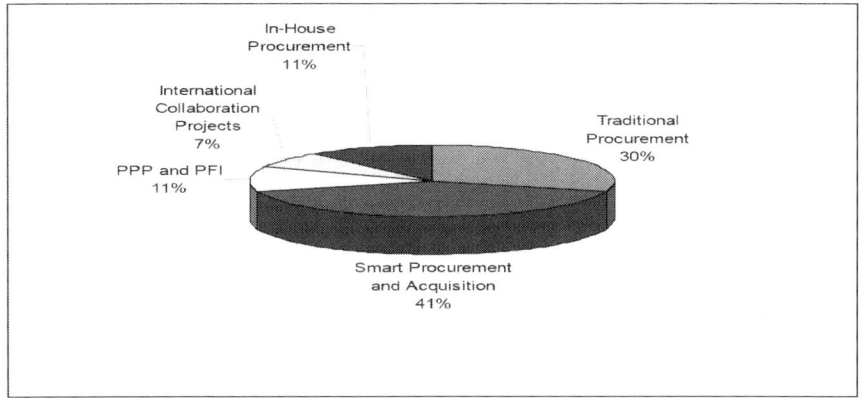

Figure 2: Procurement Route in VfM Reports

The results show that the MOD is tending more towards a Smart Procurement (introduced in SDR1998, renamed Smart Acquisition in 2000) selection process. However, the second most frequently used procurement process is shown as Traditional Procurement, this is attributed to the legacy projects (this refers to projects approved prior to the introduction of Smart Procurement). In-house procurement has become less commonly used by the MOD and in fact the descriptions of this selection process referred primarily to logistic projects and even so, these references highlighted a transition towards outsourcing in-house capacities. International Collaboration and, PPP and PFI projects are undertaken for more specific requirements, such as where interoperability is required with allies or where cost saving can be achieved through private investment. In terms of PPP and PFI Procurement, they are characterised as generic assets (such as non-combat vehicles) procured through a risk-transfer element in the contractual agreement with the prime contractor. A prime contractor is the selected supplier chosen to oversee the entire project according to the contractual agreement, with responsibilities of integrating the systems and managing the supply chain.

5.
BUYER-SUPPLIER DICHOTOMY

This investigation looked into the reasons behind the way the MOD and Industry engage; the frictions in the transactions between the two parties; and this then led to identifying how the engagement can be characterised where project success is threatened. These three investigations were the basis of the propositions put forward in the doctoral research. As already discussed above, to understand the way the MOD selects its prime contractors for a specific defence project it is important to deploy the concept of asset specificity (also known as transaction-specific investments).

The asset specificity involved in the transaction can explain the reason why the MOD has chosen a specific prime contractor, and also why it then becomes locked into that supplier for future procurements of a similar kind. This then suggests why the MOD should change the way it engages with long-term prime contractors. The legacy of the privatisation of the defence industry and the activities of these companies to consolidate their position in the defence sector has been to create monopolies in the defence sector. Even where the MOD followed a policy if liberalisation by allowing European and American defence companies to compete with British companies, the result was these companies sought mergers and acquisition to create their own monopolies in the defence sector. The result was that the MOD was faced with a supply base, which was made up of monopoly prime contractors.

The nature of defence acquisition is such that the MOD has to invest major finance and resources in the project in order for the prime contractor to be able to undertake the project. These investments create *post-contract asset specificities*, in that the prime contractor was able to use these investments to justify its continued involvement in future projects of a similar kind. Once it has the capacity to deliver a project of a specific kind, the prime contractor can make its own investments and create a *pre-contract asset specificity* to further ensure the MOD is locked-in for a long-term supply arrangement for subsequent contracts. In TCE this is known as the *Fundamental Transformation* a change from a many supplier situation to a sole supplier situation, hence the emergence of a monopoly supplier. The MOD still has its Monopsony buyer powers, as a single buyer of defence equipment in the UK, however this results in a *bilateral dependency situation*. This has created a situation where there is a balance of power (this is especially evident as UK defence companies have become dominant in the international market) in the negotiation of defence acquisition projects. There is clear evidence in the Value for Money reports, of this transformation occurring in projects such as the report on *The Construction of Nuclear Submarines in Facilities in Devonport* (HC90, 2003)shown below:

> The Department required new and upgraded facilities at Devonport for the refitting and refuelling of its nuclear- powered submarines. As part of the sale of Devonport dockyard in 1997 the Department agreed to fund the provision of these facilities which, when completed, would then be owned and operated by the new dockyard owners, DML. The provision of these facilities by the Department, together with the guaranteed submarine refitting work-stream underpinned DML's business case for its purchase of the dockyard.

It is clear from this reference that the sale of the dedicated and site specific asset of the Devonport dockyards allowed DML to attain a strategically important dockyard, while the investment from MOD provided DML with further capability to attain a long term contractual relationship with the MOD. Examples such as this demonstrate how the MOD has allowed private contractors to attain investments which they can then ensure a prime contractor position in the long-term. However, the argument put forward in the doctoral research is not to reverse this trend (as that would be highly difficult to do and undesirable due to the negative effects it may have on the defence industrial base), instead the argument is that the MOD has to reassess the way it engages with its Prime Contractors to take into account the long-term nature of the relationship.

The reason for the reassessment of the way the MOD engages with its Industrial partners is because there are clear indications that the difficulties observed in defence projects are directly linked to transaction costs in the buyer-supplier exchange. Transaction costs are additional costs to the project due to the way the transaction is undertaken and the consequence of unforeseen contingencies affecting the project. TCE outlines three environmental factors, which can affect the transaction. Of these three environmental factors *uncertainty/complexity* and *asset specificity can generate* transaction costs leading to project difficulties. *Frequency,* however, showed an inverse link to project failures, in that the frequency of the transaction can generate positive learning effects for overcoming project difficulties.

There is evidence to point to transaction costs being a part of the reason for failures to meet the value for money criteria in defence projects. In terms of uncertainty/complexity the presence of political uncertainty, technical complexity and other forms of uncertainty and complexity have a direct effect on the achievement of cost, time and performance targets. A lack of contingency planning in projects contributes to the presence of transaction costs, due to changes in the project which impact on the cost and time targets. Asset specificity factors in the project, in terms of the way the transacting partners (MOD and prime contractors) engage in adversarial negotiations where each side uses physical assets (in terms of the transfer of ownership of assets) or temporal assets (scheduling) to gain an upper hand, in the long term this creates greater costs and time delays.

The focus, thus, moves to how the MOD responds to situations where project failure occurs or is likely. TCE predicts where organisational failure occurs, i.e. the project is likely to fail due to the way the MOD and Industry are engaging, and a change of *transaction governance* is required. Transaction governance is the structural organisation of the transaction, whether the transaction occurs in the market, through internal organisation (in-house) or a hybrid (e.g. PPP). Figure 3 provides a classification of the way the MOD has responded to organisational difficulties in the Value for Money projects. Four types of responses are identified: (1) successful changes; (2) unsuccessful changes; (3) no changes; and (4) no/minimal mention regarding transaction governance.

Figure 3: Transition Types of VfM Projects

Type 1	Type 2	Type 3	Type 4
Bowman CIP	Apache Helicopters	Common New Generation Frigate	Chinook Helicopters
Future Nuclear Deterrent Capability	Defence Fixed Telecommunication System	Defence Information Infrastructure	Landing Platform for Helicopters
Helicopter Logistics	Non-Combat Vehicles for the RAF	Eurofighter 2000	The Torpedo Programme
Hercules C-130	Nuclear Submarine Facilities at Devonport		Trident Project
Logistic Support for Fast Jets			
Type 45 Destroyers			

There is an interesting trend in the spread of projects of the four types. The more recent reports published in 2002-2009 are clustered in Type 1; Type 2 is populated with projects from 1999-2003 reporting period; and Type 3 and 4 are predominantly in the 1984-1995 period. The exception is the DII project in Type 3 and Chinook Helicopters in Type 4, both being from 2008. This trend is interesting because it indicates that the MOD became more focused on transaction governance within the last decade. Its success has been in the more recent projects this may be due to the introduction of successful acquisition strategies. The trend on the whole shows that the MOD has become more willing to intervene in projects where the transaction governance is failing to deliver the desired outcomes.

The evidence also indicates that the MOD has moved to focus on partnering arrangements with its suppliers. There is, however, a clear failure in the MOD's ability to apply PFI contracting to defence projects. One of the reasons for project failure from a TCE perspective seems to be a lack of transaction governance consideration in these PFI projects. The MOD may see PFI as a useful tool to transferring greater project risk to suppliers. However, its ultimate success to achieve this aim will be in how the MOD combines this with effective transaction governance. The MOD has failed to transfer the risk to the prime contractors in the PFI projects, as has been suggested in the Value for Money reports.

There is, therefore, a clear indication that the way the MOD engages with industry has a direct link to the success of the project. The MOD is in a long term bilateral dependency with its main prime contractors in major defence projects. There are clearly environmental factors which affect the transaction, impacting the success of defence projects. Where failure is evident the MOD has in some cases shown that it can adapt to the situation by introducing new procurement policies and methods such as SMART Acquisition. The challenge is, however, in ensuring that these changes are properly supported by effective and efficient contracting processes. The following section will provide the research findings for the contractual aspect of defence acquisition.

6.
CONTRACTING PROCESS

The analysis into the contracting process focuses on three aspects of contracting: contract type, contract environment and contract negotiation. The contract types provide the references made to the contracting mechanisms used in the acquisition of the equipment or services presented in the Value for Money reports. The contract environment looks at three factors affecting the contracting process. These are the references made to the monopsony and monopoly conditions of the contracting; the human resources considerations during the contracting stage; and the frequency of the contracting between the two parties. The other aspect of the contracting process is the contract negotiations.

Passages referring to the contract negotiations were highlighted and coded, as were examples of learning gained or applied from previous experience of contracting with the supplier, and the information transfer issues, which impacted during the contract negotiations . The references made to these factors in the Value for Money reports are presented as a frequency count in the bar chart of Figure 4, alongside the Risk sub-categories (which will be discussed further in the following section).

Figure 4: Presence of Level 4 Factors in VfM Reports

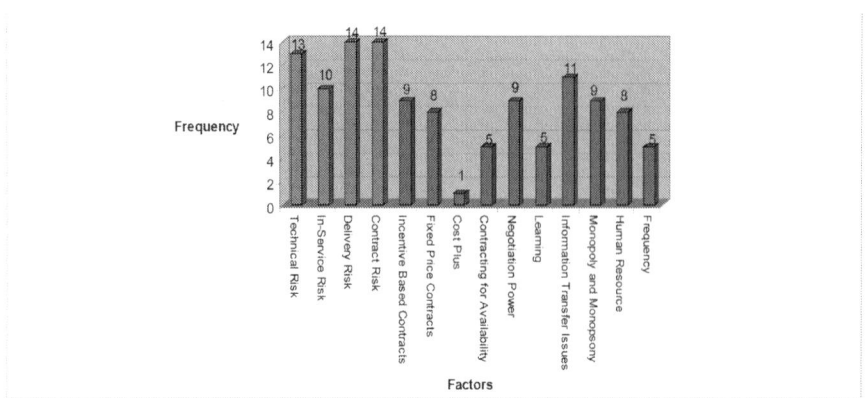

The contracting process sub-categories show a lower presence in the Value for Money reports, as compared to the Risk subcategories. This may indicate that Risk is of a greater concern in the reports, however the contracting process plays just as an important role in the achievement of project success. The interesting indication of the contract process bars is the prominence of *information transfer issues* as a factor discussed in the reports. The references to this factor provide indications where there has been a lack of communication, information disparity or inaccuracies during the contracting process, which then impact the project in terms of cost increases and/or time delays. The findings also show the frequency of the references made to the contract types, of which four are identified in the reports. The spread of these contract types are shown in Figure 5, in the form of a pie chart.

Figure 5: Contract Type in VfM Reports

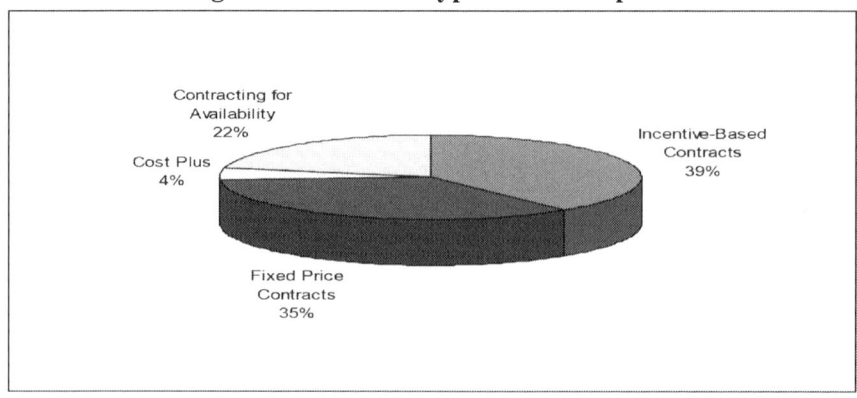

As was the case with the selection process factors, the contract types also overlap. The reason, however, is due to the fact that a defence project can have more than one contract for the various parts of the programme or due to changes made during the project. Thus, a report can refer to more than one contract type for an individual defence project or programme. The share of contract type indicates that the majority of contracts are written in the form of incentive-based contracting and fixed price contracting. Contracting for Availability has become a more popular form of contracting in projects where spare parts are required or where large orders are required in interval periods (as referenced in the *Helicopter Logistics* (HC 840, 2002)and *The Procurement of Non-Combat Vehicles for the Royal Air Force* (HC 738, 1999) respectively). Cost Plus contracting has become a thing of the past in defence contracting; its use has drastically reduced since the Levene (1987) reforms.

The analysis of the categories also looked into the relationships between the factors, in terms of the causal links. The findings showed a clear indication of a relationship between the selection process, contract type and contract risk. The relationship identifying the effect that contract risk has on the procurement route scored the highest frequency count of all the relationships, showing clear evidence that the risks identified in the contract will have a bearing on the procurement route followed. There are references to the way risk is allocated in the contract, which is dependent on the procurement route. As such, the indication would be that the risk is transferred in the contract to the supplier if a PFI procurement route is taken and if it were a SMART Acquisition project the risk would be shared by the MOD and Prime Contractor. The second highest relationship frequency was similarly the relationship between the selection process and contract type. The procurement route would outline the contract type to be followed in the acquisition. The solitary cost-plus contract is identified in the Trident project, which followed a traditional procurement route, a majority of the fixed-price and incentive-based contracts were from SMART Procurement and Acquisition projects, and the Contracting for Availability contracts were identified in the PFI projects. These links show that the procurement routes will determine, predominantly, the contract types used in the acquisition.

Table 2 provides a more representative picture of contract types used by the MOD over the previous three decades; the table is compiled from DASA reports[1]. More than half of defence contracts in the 1980s were priced with reference to market forces. This trend continues in the 1990s, although during this period the share by value of competitive pricing contracts had doubled. The number of contracts priced using market forces had reduced to between a third and two thirds; this is mainly due to an increase in competitive pricing contracts. In the 2000s, competitive pricing became the dominant contract type by value and share. The MOD has awarded more incentive-based contracts in its more recent projects. The MOD has followed this model because it believes it's the best way it can link payment to deliverables in order to align its interests with that of the supplier. Using mechanisms such as key performance indicators the MOD attempts to ensure that the supplier is delivering key user requirements. The levels shown for the incentive-based contracting may differ in Table 2 as compared to the share shown in Figure 5. However, this is more of a definitional factor rather than disparity, since it is not clear how contracts let according to competition or market forces are designed, they too might have an element of incentives.

Contract Type (Contract priced by)	1980s		1990s		2000s	
	Value (% share)	Number (% share)	Value (% share)	Number (% share)	Value (% share)	Number (% share)
Competition	22 – 44	12 – 14	41 – 71	16 – 54	51 – 70	34 – 42
Market Force	14 – 26	59 – 72	4 – 23	33 – 67	8 – 12	25 – 41
Estimates	20 – 35	8 – 16	16 – 30	12 – 17	18 – 30	17 – 28
Incentives to minimise costs	3 – 20	1	1 – 5	0 – 1	2 – 17	0 – 3
Cost-Plus a % Fee	4 – 22	2 – 11	1 – 4	0 – 2	0	0

Table 2: MOD Contract Types 1980-2010 (DASA (1992,…, 2010))

The analysis of the contract negotiations using a TCE perspective provides some interesting findings on the way the MOD and the prime negotiate in the bilateral dependency situation. In the balance between the buyer's maximum price and the seller's minimum supply price, uncertainty and asset specificity will play a role in which side gets to achieve the optimal outcome. In the data analysis the evidence points to the MOD being most affected by

---

1   Table 2: MOD Contract Types 1980-2010 (DASA (1992- 2010))

the uncertainty and asset specificity of the transaction, it is usually the MOD which moves towards its maximum price rather than the Primes moving to the minimum price. The Value for Money reports and the Major Projects Reports indicate that the MOD usually end up paying more than anticipated.

The MOD has attempted to apply certain mechanisms in the contracting process to redress the balance of power in contract negotiations, such the use of competition, contractor shadowing and requirement changes, in order to reduce monopolistic powers. These mechanisms have created an adversarial relationship with Industry, rather than achieving efficiencies. The main reason for this is that the MOD has focused on creating competitive pressures on Industry in order to achieve competitive pricing, rather than realising that the main reasons for project difficulties are the presence of uncertainty, complexity and risks in projects.

7.
UNCERTAINTY, COMPLEXITY AND RISK AND THE MANIFESTATION OF OPTIMISM BIAS

As has been suggested throughout the report there is a significant role played by the factors uncertainty and complexity, and risk in defence project difficulties. Uncertainty and complexity are identified as a singular category as they have the same effect on the project. However whereas uncertainty is regarded as a condition of the unknown future due to various reasons (economic, political or technological), the conditions of complexity take on a more technical manner in that it is the challenges to knowledge. For example, uncertainty is represented by events such as the political uncertainty in the international collaboration project of the *Eurofighter 2000* (HC 724, 1995). Whereas complexity is in terms of the inherent technical complexity of the requirements such as in the *Delivering Digital Tactical Communications through the Bowman CIP Programme* (HC 1050, 2006) report. Uncertainty and complexity have no sub-categories due to the homogeneity of the consequence of the causes, risk is characterised as four kinds in the Value for Money reports. These four kinds are (as shown in Figure 5) technical risk, delivery risk, in-service risk and contract risk, in any individual project one or more risks may affect the project.

Changes to user requirements are one of the most common reasons for uncertainties in defence contracting. The MOD has to make a number of changes to its requirement, from project conception to execution, due to a number of reasons such as political, policy, budget, technical, and security changes amongst other reasons. The supplier can find it difficult to adapt to the changes and this can have consequences in budgeting, scheduling and performance. As well as requirement changes, technical factors have been identified as a major reason for difficulties in defence acquisition projects in the Major Project Reports. In *The United Kingdom's Future Nuclear Deterrent Capability* (HC 1115, 2008) report there are references to the impact of technical factors on the project. It is mainly due to the inability of the transacting parties' to foresee or create contingencies for uncertainties in the technical challenge which they may face, that makes project control difficult in defence.

The impact of uncertainty is in the planning stage where unforeseen consequences cannot be predicted. Although it manifests in the delivery stage, where the schedules, budgets and resources are prescribed, uncertainty can have a major destabilising effect on these factors. In some cases, such as *Providing Anti-Air Warfare Capability: The Type 45 Destroyer (*HC 295, 2009) project, the planning can be overly optimistic and the project starts badly and continues to deteriorate, until corrections can be made.

Gardener and Moffat (2008) highlight in their study of defence acquisition using a game theory approach, that there exists an optimism-bias in defence projects. They show that the presence of uncertainty creates a situation where the MOD follows an optimistic rather than a realistic strategy, when planning spending, schedules and performance requirements. Uncertainty plays an important role in the way the transacting parties behave. As they suggest, when there are higher levels of uncertainty the parties feel they will gain from acting opportunistically and creating an optimism-bias environment. This optimism-bias then distorts the cost, time and user requirement targets for a defence project.

While game theory can provide a sense of understanding for the motivations of the optimism-bias and what the optimal choice in the game should be; it provides little in terms of a solution to the problem. In that, these games are given as pre-determined moves rather than changeable situations. The doctoral research instead approaches UK defence acquisition with the view that by identifying characteristics of the acquisition process, recommendations can be made to create a solution to the major challenges of defence acquisition. The solution to optimism-bias, in my view, must be in the way the MOD fully implements the 'open relationship' principle of SMART Acquisition, and seeks greater involvement from the prime on what is feasible and deliverable. Rather than the current situation where the MOD is told what it wants to hear from Industry, as was intimated by an interviewee in the study by Kebede *et al.* (2009).

Figure 6: A Descriptive Model of Defence Projects Time-Cost Progression, based on MPR studies

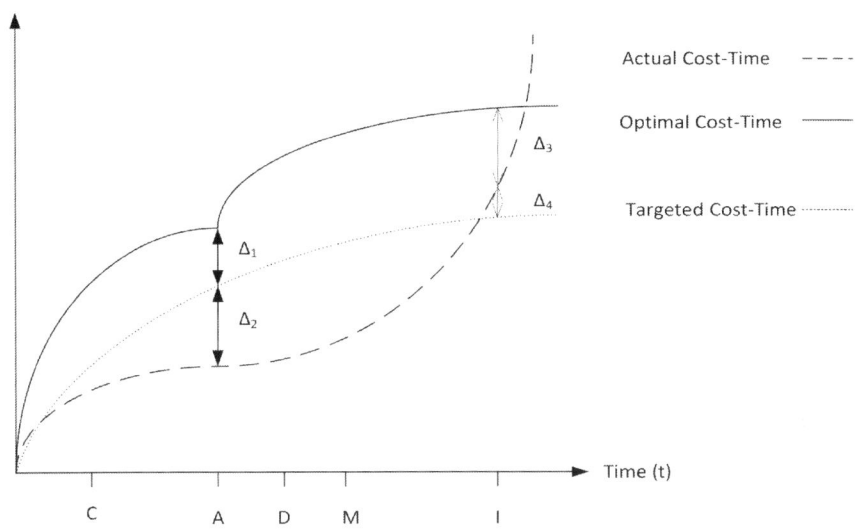

In Figure 6, the intention is to capture, very generically, the findings of the Major Projects Reports and the literature review of UK defence acquisition from the thesis. I intend to show two aspects: first, the way cost progresses over time for major projects, from the concept phase to the in-service phase (CADMI-(D): the disposal phase is out of the time range of the graph), in three scenarios, actual, optimal and targeted. Second, the aim is to show the differences (Δn) in investment (represented by cost gaps) at the assessment phase and in-service phase. The three lines show very different trend lines.

The progression of the targeted cost line represents the optimism-bias effecting defence project planning, in that the decision-makers underestimate the investment required. While optimism-bias may be one of many reasons (such as project errors and exogenous factors) for underestimation, it is one of the most identifiable and removable factors. The line represents a steady cost progression throughout the project life cycle. The actual cost line illustrates a rather cubic expression, with the funds for the defence project starting steadily and at the turning point (the assessment phase) the cost escalating, due to the lack of investment during the concept and assessment phases. Project costs continue to escalate at the in-service phase, due to a variety of reasons, such as user requirement changes, technical difficulties and so on. The cost is expected to level off as it approaches the disposal phase; however this is not within the timeline of the graph.

The optimal cost progression illustrates the cost trend line for how a project could be planned in a two-step investment plan. It shows greater investment during the concept phase (more funds for risk reduction and technical development work). An injection of funds (at a lower rate than the concept phase) is then expected in the second step at the assessment phase. This will ensure investment is available for the project to meet its in-service date and user requirements, resulting in the cost levelling out at the in-service phase. Greater cash flow is required in this investment plan; it is however the optimal cost plan with respect to the desired outcome being achieved in time and performance.

The differences in funding at two points in the acquisition process, the assessment and the in-service phase, are shown. At the assessment phase there are two interesting investment differences:

1. The difference between the optimal cost and the targeted cost, $\Delta 1$, shows a lack of investment into defining and planning defence projects, which then impacts project costs at the assessment phase (described in Chapter 5 of the thesis).

2. The difference between the targeted cost and actual cost, $\Delta 2$, demonstrates the optimistic expectation, in investment, for effective preparation to progress to the demonstration and manufacturing phases. The difference between the targeted and optimal cost represents the total investment gap during the assessment phase.

The cost gaps at the in-service phase represent the expected costs when the project reaches its in-service date:

3. The cost of the project when following the optimal trend line is higher than that of the actual cost, as shown by $\Delta 3$. However, the optimal trend line levels off whereas the actual cost trend line continues to grow and surpassing the optimal line over time. This is because in the current situation unforeseen costs impact the project (identified as partly being of a transactional nature in this research).

4. The cost of the project in the targeted cost line is optimistically planned to be lower than the actual cost. While $\Delta 4$ is smaller than $\Delta 3$, the actual cost is still rising whereas the optimistic expectation is that it will level off. The optimal line shows that the cost can level off, but at a higher cost (due to greater investment) than the optimistic view shown by the target trend line.

While the cost of the project is higher in the optimal cost trend line it delivers a better overall cost than the actual progression of current projects. The optimal cost line takes into account two important principles of SMART Acquisition: early investment and whole-life costing. It is also expected that with experience (transaction-specific learning) of the optimal cost plans, the timeline would improve with shorter gaps for the CADMI(D) timeline.

Other than the presence of uncertainty and complexity, and the corresponding effects of optimism-bias on the project, there are risks within the project that must be identified, evaluated and reduced. Four project risk types are identified from the Value for Money reports: contract risk, technical risk, delivery risk and in-service risk. Contract risk is in the form of an ineffectual contract agreement or the presence of a risk which can impact the contractual agreement. This risk type can affect every aspect of the project. Technical risk is most evident in the development and production stage of the project and is mainly due to unforeseen technical difficulties (due to the complexity of the systems) or ineffective risk reduction work during the assessment phase. Delivery risk is the risk of missing in-service target dates as a result of delays in the project arising from contract or technical risk. In-service risk is as the name suggests the risk to the project at the in-service phase and is a result of the combination of all the other risks, operational challenges or renegotiations at the support phase.

The first step in risk mitigation is in the type of risk agreed under contract agreements. Different contract types require differing levels of risk management by the MOD. Contract agreements can place the risk on the MOD, or on the prime contractor, or share the risk. Where the risk is taken on by the prime contractor, they will charge a premium to cover the project risk in the form of a fixed price contract. In some cases the MOD are willing to absorb the risk. In *Transforming Logistics Support for Fast Jets* (HC 825, 2007) in the availability contract for the Tornado support contract the MOD took a greater degree of the risk due to its satisfaction with the partnership agreement and the pre-contract work undertaken. Where there is a lack of risk reduction work and effective partnering arrangements, contract risk can be increased by an inefficient contract agreement.

The prime is reluctant to take on risks in the contract and if it does so it will aim to cover it by charging a premium. Therefore, risk becomes an important element in the way partnership succeeds in the project. The findings of the investigation into project transition types, shown in Figure 3, demonstrates organisational failures were predominantly due to a lack of risk management in the contract (mainly in PFI procurements). A key step to reducing the project risk is to adopt an effective partnership approach with the prime contractor. An NAO report (HC 1047, 2006) points out 'contracts are mostly likely to support successful project delivery if they are negotiated against a common information base and with an understanding of stakeholders' aspirations. In applying the contract it will be important that all parties understand the way in which achievement of the desired outcomes is incentivised and the position of each stakeholder is protected'.

The transition to effective partnership arrangements has enabled the MOD to partner with its suppliers and deliver improved procurement and support contracts. Thus, in delivering effective risk management the projects referenced as being successful mention 'co-located teams' and 'communication' as key to working practices. The strengthening of the relationship between the MOD and prime, emphasised the creation of IPTs have been an important step in improving risk management in defence projects. An appreciation of technical risk in project development and production is a key factor to project success. A common reason for project cost and time variance discussed in the Major Project Reports was identified as technical factors. The reasons given for the continual presence of technical factors in the Major Project Reports are due to a lack of risk reduction work and investment in the Assessment Phase. Defence projects involve a high level of technological complexity, where the prime contractor is responsible for the integration of systems of systems.

Technical risk is lowered where the procurement is of a standardised product with non-specific investment required in physical assets. However, where asset specificity is high, the technical risk may increase. The report on *Delivering Digital Tactical Communications through the Bowman CIP Programme* (HC 1050), highlights the challenges which faced General Dynamics. However, it is an example in which the prime contractor seemed to invest time and money in order to deliver the customer requirements. The evidence in the Major Project Reports and the Value for Money report shows that Bowman CIP under the partnership of the MOD and General Dynamics was a successful project (with time and cost increases at a mere 5 per cent). While General Dynamics inherited a lot of the technical difficulties from Archer's Consortium, they worked to effectively reduce the technical risk.

Delivery risks can affect the scheduling of the project and can have a negative impact on in-service dates. The MOD and its industrial partners have to work in partnership to reduce the impact of delivery risks. The rationale behind investment in the early stage of the project of the Assessment Phase is to reduce the risk of in-service delays as well as cost increases. Delivery risk is based on unrealistic in-service dates, which impact all aspects of the project. Time pressures can reduce the amount of time spent during the Assessment Phase which impacts the effective delivery of the project. The risks in the in-service phase highlight the need to continue the partnership with industry post-procurement. This signifies the need for through-life management of defence projects 'from cradle to grave'.

The risks involved in the in-service phase are linked mainly to the operational requirements of projects and are significant to the customer, the Armed Forces. As explained by NAO (2003) report 'through enhanced planning, Through-Life Management should lead to improvements in the affordability of future programmes, a reduction in the number of 'surprises' encountered by a project and better delivery of integrated military capability rather than individual items of equipment'. The NAO are pointing to the importance of planning for future costs, an important principle of SMART Acquisition: 'a whole-life approach, typified by applying through life costing techniques'. However, their focus goes beyond just the costing of individual platforms, but of integrated projects with funding for encompassing new technologies and capabilities such as Network Enabled Capability.

8.
RELATIONAL CONTRACTING

The introduction of the IPT mechanism has been one of the most tangible acquisition policies taken by the MOD to foster a close working environment with industry. It is one of the seven Smart Acquisition principles and its implementation was fast tracked by the SMART Procurement Implementation Team (NAO, 2002). Defence Equipment and Services (DE&S) described it as the primary building blocks of the organisation (MOD, 2007). The aim of the IPT is to create the collaborative environment and partnering the MOD wishes to foster with industry. It is therefore an ideal candidate to test the successful realisation of a bilateral governance framework.

Figure 7 represents the impact of the factors identified in the Value for Money reports on the IPT mechanism, and it provides the framework in which it can be understood in the context of the doctoral research. There are three elements to the Venn diagram. The top circle represents the structure of engagement between the MOD and the prime in the IPT. There is a circle describing the positive impact of defence acquisition factors on the IPT and next to it, the negative effects. The boxes explain the effects that the defence acquisition process factors have on the IPT mechanism. Each box corresponds to an overlapping area of the circles. The impacts are to strengthen the IPT, to weaken the IPT and to neutralise the effect on the IPT.

Figure 7: Venn Diagram of IPT Mechanism

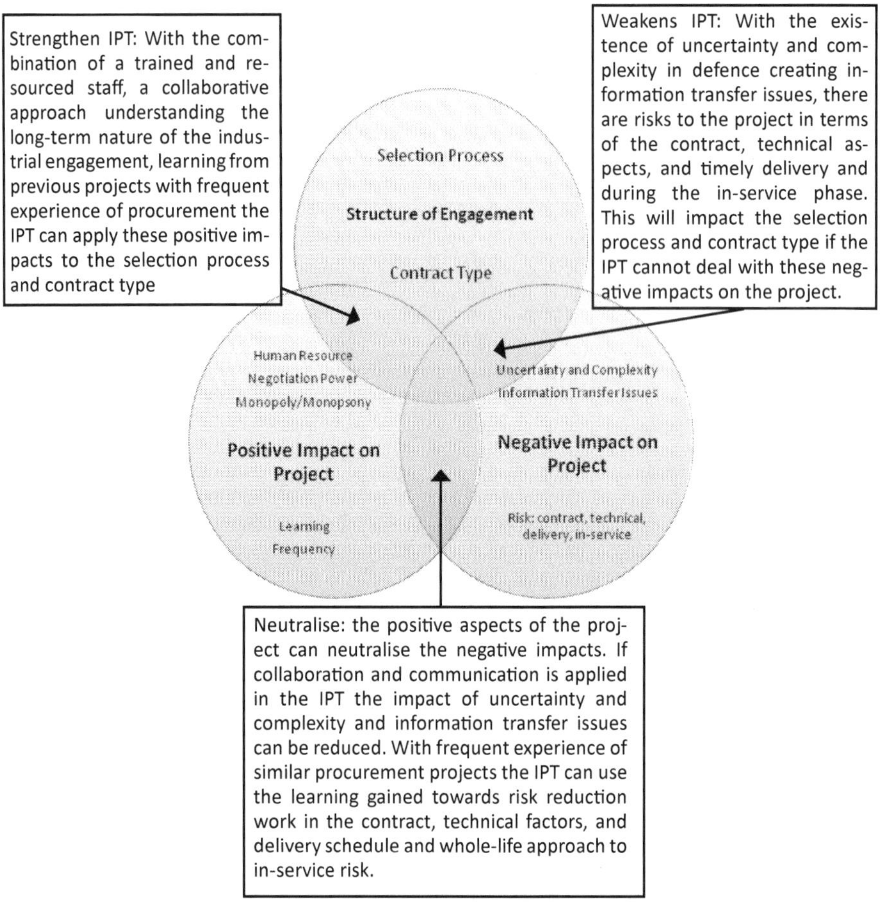

Strengthen IPT: With the combination of a trained and resourced staff, a collaborative approach understanding the long-term nature of the industrial engagement, learning from previous projects with frequent experience of procurement the IPT can apply these positive impacts to the selection process and contract type

Weakens IPT: With the existence of uncertainty and complexity in defence creating information transfer issues, there are risks to the project in terms of the contract, technical aspects, and timely delivery and during the in-service phase. This will impact the selection process and contract type if the IPT cannot deal with these negative impacts on the project.

Selection Process
Structure of Engagement
Contract Type

Human Resource
Negotiation Power
Monopoly/Monopsony

Uncertainty and Complexity
Information Transfer Issues

Positive Impact on Project

Negative Impact on Project

Learning
Frequency

Risk: contract, technical, delivery, in-service

Neutralise: the positive aspects of the project can neutralise the negative impacts. If collaboration and communication is applied in the IPT the impact of uncertainty and complexity and information transfer issues can be reduced. With frequent experience of similar procurement projects the IPT can use the learning gained towards risk reduction work in the contract, technical factors, and delivery schedule and whole-life approach to in-service risk.

The structure of the engagement in the IPT is defined by the selection process and contract type. When the MOD and the prime make relation-specific investments, it can improve the engagement of each party in the IPT. However, if the parties do not invest in the collaborative relationship, there can be a weakening of the IPT. This is explained in a case study of an Advanced Military Vehicle IPT by Kebede *et al.* (2009):

> One important idea behind the IPT was to have a co-located project team that could take advantage of team working, good communication and cooperation. This notion was to be realized through continuous interaction enabled by prescribed tasks, teams and routines. The management of relationships (customer, user, suppliers) was therefore a key tactic in this programme. However, the original bid was won through competitive tendering, and the

dynamics inherent in such a process have placed significant constraints upon the IPT's ability to collaborate, because of the constraints of a project won by competition, with "a very tight contract, budget and margins squeezed to win the work". For example, the proposed shared data environment (SDE), which would enable improved communication and coordination between industry and customer, was not effectively implemented due to cost and commercial constraints'.

This citation provides some interesting insight into the fundamental difficulties in the implementation of the partnering approach the MOD is aiming to follow. The procurement route chosen in this project is clearly a Smart Acquisition route, with emphasis on collaboration through the IPT mechanism; however the contractor selection process was followed a competitive tendering approach. This has clearly created conflicting outcomes, whereas the mechanisms for a partnership approach are available in the form of the IPT mechanism, the tensions created by the competitive tendering process has reduced, if not nullified, the scope for partnering. There is, therefore, a clear indication that the procurement process needs to be cohesive with the contractor selection approach if the aims of the contractor engagement are to be realised. Hence, one of the recommendations made in this research is that the choice of a partnership approach needs to be followed using a long-term approach with a preferred prime contractor. The decision has to be between either a partnership route or competitive tender route, an attempt to apply competitive pricing pressures on the bidders will lead to transaction costs and failures to meet targets further in the project life cycle.

The selection process and the contract type are seen to have knock-on effects on the IPT mechanism. A lack of investment in the shared data environment may increase the presence of negative factors, such as uncertainty and complexity, creating information transfer issues, which may lead to risks in the project in terms of contract, technical, delivery and in-service risks. The use of IPTs for support contracts were referenced in the *Helicopter Logistics* (HC 840, 2002) report and have been identified as a successful change to governance arrangements. However, there were also some concerns that not enough time and resources were provided to implement new working practices successfully. IPTs can be a positive influence in the project management; however, there is a clear need to provide the appropriate resources to strengthen the IPT mechanism. The NAO have in *Transforming Logistics Support for Fast Jets* (HC 825, 2007) praised the effective use of the IPT mechanism for Fast Jets support contracts. The IPTs have followed 'open book accounting and sharing of information and benefits between projects'. Where relational contracting is utilised in the form of relation-specific investment in communication and trust, the bilateral governance can be seen as a successful transaction governance, as exemplified by the Fast Jets support contracts. However, where the IPT fails to engender the principles of IPT engagement, then the transaction costs increase and organisational failure can seem more likely.

To ensure the successful application of the IPT mechanism, there needs to be a congruent contractual process that complements the IPT ethos. Relational contracting is the contracting process best suited to engendering the closer and open nature of defence acquisition under the IPT mechanism. While the contract type may continue to be under a fixed price or incentive-based contract, mainly the latter, the contract would follow a relational contracting agreement. In relational contracting, there is a move from prescriptive contracting to one that is more flexible and adaptable, rejecting the classical or neo-classical approach where ultimate dispute resolution is decided in the courts. Transacting parties work together to create their own conflict resolution mechanism, mainly in the form of joint-problem solving exercises using social processes such as trust and communication.

The MOD hopes to achieve this by replacing its traditionally adversarial relationship with industry to one with a more open and collaborative nature. The transition towards relationally-governed exchanges through social processes may take time, but it is aided by the use of SMART Acquisition and the IPT structure. There will be resistance from each side to sharing information which they view are sensitive to their bargaining position, however once relational contracting is supported by improved trust between the parties such issues may be less problematic. A relational contracting approach to IPTs may provide greater freedom for them to work more effectively, with less contractual threat hanging over the MoD and industry joint team. It would remove the perception of IPTs as a monitoring mechanism intimated by Cullen and Hickman (2001), creating a full recognition of IPTs as drivers of joint-working arrangements for the MOD and industry. It would create more opportunity for DE&S staff to examine ways of maximising the IPT mechanism, rather than spending a great deal of effort and funds on legal advice on contracts.

The MOD (2005) comments 'the emphasis on our future approach to ensuring value for money has highlighted the need to place greater emphasis on fostering better, and where appropriate, longer term relationships with our key suppliers, and the use of appropriate commercial tools, including competition or formal partnering agreements. This must be underpinned by greater openness and transparency, with a common and more explicit understanding of how to achieve best value for both Defence and industry'. Competition may create the opposite of what is desired in a partnering agreement, as pointed out in the AMV IPT case study, competition creates strains in the IPT mechanism which reduces its efficacy. The MOD may have to view the procurement process as a 'competition V partnership' choice prior to tender. While, competition can provide advantages in pricing and aligning the Prime to the MOD's interest. Partnership can reduce the long-term transaction costs, by reducing the project risks and creating a more collaborative and open relationship with the Prime. In the end, the MOD may decide which procurement route it prefers on a project-by-project basis. The tensions in the relationship, however, occurs when the MOD decides to follow a partnership route but uses competitive bargaining prior to this, as exemplified in the AMV case study.

10.
CONCLUSION

The conclusion of this research is for the MOD and Industry to enter into a new era of acquisition policy characterised by strong and effective *long-term relationships*. Some may see this as suggesting a return to the days of the cosy relationship; however this would be to misunderstand the advances made in defence acquisition policy over the last few decades. The balance of power in the defence sector is moving towards a more level playing field as multinational defence companies increase their presence internationally and diversify their customer base; however the domestic buyer can still rely on its regulatory powers and export policies. There is more to gain through working in collaboration than adversarial engagement. The basis of a relational contracting approach is to foster an ethos of communication, collaboration and cooperation. It is also important under this approach that the MOD and industry enter flexible contractual agreements, which provide space for unforeseen circumstances. The contractual agreements of current defence transactions are too rigid and make it difficult to respond to changes without having to renegotiate the contract.

The solution to such difficulties is to allow the IPT to resolve contractual disagreements. Since contractual disagreements are usually based on technical factors or exchange hazards (IPR and so on), the IPT is best placed to resolve issues of technical difficulties or trust issues. The IPT mechanism is important not only due to the co-location of team, but also the closer working relationships that it fosters. IPTs are also best placed to implement transaction-specific learning in order to learn from retrospective studies of project risks and reduce the impact of uncertainties on the project.

Incentive-based contracts have the best structure to provide flexibility in contracting. However rather than creating penalties where milestones and targets are missed there should be an opportunity for the parties to jointly tackle these issues by discussing ways to remove these delivery risks through trade-offs, investment in monetary or human capital, or bringing in a third party to provide support (technical, consultancy or logistics). Fixed/Firm price contracting can be applied to support contracts where there are less contractual hazards. These should however be contracted not at the procurement phase, but as the partners move towards the in-service phase. With relational contracting, it is less likely that the hold-up problem would impact the support contract since there is a long-term objective in the relationship.

The ultimate contractual safeguard in relational contracting is reputation. Long-term partnering provides the defence companies with risk minimisation, increasing their ability to survive in the future. They would be risking their future survival if they were to act opportunistically for short-term gains. The MOD might reach a point where it would be willing to look at the defence supply chain to promote Small to Medium Enterprises (SME) acting as second-level suppliers, where it feels there is irreparable damage to the trust with the long-term partner. However if it wants to have this possibility, it needs to improve its industrial strategy, with respect to the role SME's play in the defence market. In turn the MOD would also have an interest in ensuring it does not endanger the partnering approach by acting opportunistically. If the MOD were to act in such a way, this would return the relationship back to an adversarial one, in which contract disputes and project failures are the most likely outcome. It will have to therefore reassess its use of some mechanisms it has used in the past, to gain competitive pricing such pseudo-competitions, contractor shadowing and unnecessary requirement changes.

The MOD should not abandon all other procurement routes in this new policy of relational contracting; it should employ PPP arrangements such as PFI (though it needs to design better governance arrangements for this procurement type) and international collaborations (with NATO or EU allies) where necessary. Where the MOD wishes to transfer greater risk in specific projects (PFI), or where allied capability would benefit from joint-capability (international collaboration), it may be appropriate to use a separate procurement route to partnering. However, it is the recommendation of this research that in most of its national acquisition, where the MOD needs to engage bilaterally with industry, the MOD should follow a relational contracting approach.

RECOMMENDATIONS:

Based on the analysis, the following recommendations are put forward, though further research would be required to make more specific recommendations of changes at the process level:

- **The Relationship:** *entering relational contracting arrangements based on a long-term partnering approach with Primes. The MoD needs to create a new industrial strategy, which will assess how this new approach will be embedded as part of the defence acquisition process. This could be addressed in the long awaited follow-up to the Defence Industrial Strategy White Paper. This will look at how prime contracts are identified, selected and delivered. It would also be an effective medium to communicate the MoD's desire and commitment to a new approach to industrial relations.*

- **The Contract:** *designing and introducing flexible incentive-based targets. The NAO and MoD (with contributions from Industry) should work together to identify mechanisms which would improve the flexibility of incentive-based contracts. This should focus on how targets and milestones are assigned to contracts, and how the MoD can encourage industry to be more open in their technical capacity.*

- **Conflict Resolution:** *removing penalties and replacing them with joint-problem solving mechanisms which may use third party expertise, in agreement with primes, where targets are being missed. Third party involvement such as consultancies, technical experts, and logistic support can provide a useful service in dealing with difficulties in major defence projects. By removing penalties and replacing them with these problem-solving mechanisms, the focus turns to solutions rather than litigation.*

- **The Mechanism:** *continuing to improve and promote the IPT mechanism. The IPT Mechanism should be supported by having high-level personnel involved from the MoD and Industry. Project managers need to be empowered in their decision-making and management of the projects.*

- **The People:** *investing in training and developing DE&S staff, which are at the frontline of acquisition policy and practice. If the MoD is to realise its aim of becoming an 'intelligent customer', then it needs to invest in its negotiating and technical capacity, especially in its staff.*

- **Database of Learning:** *creating an acquisition database, which IPT members can access to gain transaction-specific learning. Some restrictions may have to be in place where there may be IPR conflicts. How-*

ever, this database can be used to learn lessons from past procurement experiences.

REFERENCES

Bishop, P., (1995). Government Policy and the Restructuring of the UK Defence Industry, *The Political Quarterly,* 66(2),174-183.

Gardener, T., and Moffat, J., (2008). Changing Behaviours in Defence Acquisition: a Game Theory Approach, *Journal of the Operational Research Society*, 59(2), 225-230.

Humphries, A.S., and Wilding, R., (2001). Partnership in UK Defence Procurement, *International Journal of Logistics Management,* 12(1), 83-96.

Kebede, E., Lowe, D., Maytorena, E., and Winch, G.M. (2009) UK Defence Acquisition Process for NEC: Transaction Governance within an Integrated Project Team. In NPS(Naval Postgraduate School), 6th Annual Acquisition Research Symposium: Acquisition Research: Creating Synergy for Informed Change, May 13 -14, Monterey, California.

Kebede, E., (2011). The Application of Transaction Cost Economics to UK Defence Acquisition, The University of Manchester, Unpublished Thesis.

Levene, P., (1987). Competition and Collaboration: UK Defence Procurement Policy, Lecture given to RUSI: London.

McIntosh, M., (1993). Defence Procurement Policy: The Way Ahead, pp. 24-35 in Whitehall Paper Series ed., Defence Procurement: Trends and Developments. RUSI: London.

MOD (2005) Defence White Paper: Defence Industrial Strategy. Cm 6697, London.

MOD (2007) Defence Equipment and Support: Establishing an Integrated Defence Procurement and Support Organisation. London.

MOD (2011) Defence reform: Report into the Structure and Management of the Ministry of Defence. London.

NAO (2002) Implementation of Integrated Project Teams. HC 671 Session 2001-2002.Ministry of Defence: London.

NAO (2003) Through-Life Management. HC 698 Session 2002-2003. Ministry of Defence: London.

Williamson, O.E., (1975). Markets and Hierarchies: Analysis and Antitrust Implications. New York, Free Press

Williamson, O.E., (1985). The Economic Institutions of Capitalism, New York, Free Press

Winch, G.M., (1989). The Construction Firm and the Construction Project: A Transaction Cost Approach, *Construction Management and Economics,* 7, 331-345.

PART 2:

APPROPRIATNESS AND *APPLICABILITY* OF ICT FOR GROWTH AND DEVELOPMENT

CHAPTER 9:
Assessment of Knowledge Management Practices in Jimma University: Consideration of Technology, Leadership, Organization and Learning Pillars

Haftamu Ebuy, Rahel Bekele and Worku Jimma

1
INTRODUCTION

In today's knowledge based economy, there is a dire need for modern organizations to integrate Knowledge Management (KM) practice in their organization process and structure in order to extend their success and values for sustainable organizational development and competency. Organizations are highly investing on organizing and use of their intellectual capital (David, 1997).

This is due the reason that utilizing the organizational knowledge determines the success and maintains competitive advantage of a given organization. In this regard, Knowledge Management (KM) is created as a tool for this purpose (Senge, 1990).

According to Theriou *et al* (2010) KM has been a natural evolution over the early years of the twenty-first century, and a hot topic in several business communities. The ability to manage knowledge has become increasingly more crucial in today's knowledge economy. Jones (2003) pointed out that KM is an integrated, systematic approach to identify, manage, and share all of the department's information and knowledge assets, including databases, documents, policies and procedures, as well as previously unarticulated expertise and experience resident in individual officers.

Besides, Theriou *et al* (2010) mentioned that KM enables the existing individual knowledge to be captured and transformed into organizational knowledge, which in turn should be diffused and shared among many employees. KM is based on the idea that an organization's most valuable resource is the knowledge of its people. This means that creating, sharing and using knowledge are among the most important activities of nearly every person in every organization (Servin, 2005).

According to Yang *et al* (2011), the tangible assets like land, labour and capital are no longer sufficient to evaluate the real value of an organization's effectiveness and efficiency rather the efficiency of using the intangible assets of the organization that is knowledge are reinforced to identify the value of an organization. Akmar and Lee (2004) stated that nowadays, people are aware of the importance of knowledge and ways to acquire, recognize, capture, retrieve, use or measure, manage and collaborate knowledge, so that knowledge can be shared without losing it.

Managing organizational knowledge assets can be more effective if key elements of the organization are well integrated and involve and contribute to the enterprise practice of KM. Stankosky (2005) mentioned the successful and integrated KM enterprise learning is a function of technology, learning, organization and Leadership pillars. Also Park (2005) supported, managing an enterprise's knowledge assets can be more effectively achieved by creating KM programs using a defined framework of key elements (*i.e.,* the four pillars of KM). The four pillars clearly provide fundamental architecture. KM requires the integration and balancing of leadership, organization, learning and technology in an enterprise-wide setting.

In the advancement of knowledge and KM, universities are no exception. Mikulecka, & Mikulecky (2000) pointed out that from the learning perspective and mission of knowledge processes, university environment seems suitable to KM practices and opportunities to operate and benefit from KM. Its nature, vision, mission, objective and intellectual capital (*i.e.,* human, customer, structure and intellectual property) seems suitable to KM practices. Modernization of higher education has forced the institution to store, manage and use existing information and knowledge store in a better way in order to meet new accountability, effectiveness and efficiency requirement (Pircher & Pausits, 2011). Lubega *et al* (2011) in their study of KM Technologies and Higher Education Processes stated that "To achieve their goals, higher education institutions must try to align operational processes and organizational learning with KM technology solutions to create a performance improvement environment that strategically leverages KM technologies with higher education processes".

With respect to competitive advantage of knowledge in universities, According to Anvari *et al* (2011), beside establishment of innovation and consequently creating new knowledge, academic institutions need to identify and use the existing intellectual capital systematically through proper KM approach. The main aim of the present study was to assess the Knowledge Management practices in Jimma University using the four KM pillars: learning, leadership organization and technology and provide the guidance to improving KM practices in the university.

2.
METHODOLOGY

Survey study designs with both quantitative and qualitative method of data collection was employed. The assessment model was designed to cover four key areas of KM practice indicators; learning, organization, technology and leadership. Descriptive statistical methods such as mean and standard deviation were deployed. Also, the t-test and correlation coefficients was applied to compare the mean of the sample and to investigate correlations.

2.1 Participants of the Study

The study population was selected through proportionate stratified random sampling by grouping the source population into academic and non-academic staff. Questionnaires were distributed to 364 participants. Out of this 331 (90.9 %) and 164 (97.6%) from academic and 167(85.2%) respectively returned the questionnaire and used for analysis.

Category Specification	Frequencies (%)
Job	
Academic	164 (49.5)
Non-academic	167 (50.5)

Sex	
Male	245 (74.0)
Female	86 (26.0)
Experience (yrs)	
1-5	220 (66.5)
5-10	98 (29.6)
>10	13 (3.9)
Educational qualification	
Diploma	41
Degree	128
Second degree (MA/MS.c)	151
Third degree (PhD)	11

Table 2.1 shows socio-demographic characters of the respondents.

2.2 Data Collection Method and Instrument

The study employed a structured questionnaire having 53 questions using a five-point Likert scale responses ranging from 1=strongly disagree to 5=strongly agree (1 = strongly disagree, 2 = disagree, 3 = neutral, 4 = agree, 5 = strongly agree) framed to collect responses. The survey composed to measure four area of KM practices adapted from "Knowledge Management Assessment Tool (KMAT)" (Martha, 1998) and "The Know-all 10: A quick Knowledge Management Assessment" Skyrme, (1999). Also, participants from both groups who were not included in quantitative study were considered to participate on in-depth interview. The results in Table 2.2 show the Cronbach coefficient for all the variables in the model were above the critical value of 0.7 (Sekaran, 2000).

Pillars	No of items	Cronbach,s alpha
Learning	18	0.91
Technology	10	0.85
Organization	11	0.89
Leadership	14	0.88
KM total	53	0.96

Table 2.2 Cronbach Alpha Instruments in Each Pillar

3. RESULTS

3.1 Level of Knowledge Management Practice

Table 3.1 shows the perceptions of respondents in the level of KM practices in Jimma University and the four KM indicators. The level of KM practices was assessed by the mean average value of the indicators. The result obtained revealed that the mean value and ranges of standard deviations of academic and non-academic participants vary and there is a significance difference in the perceptions of KM practice in the university.

Table 3.1 level of KM practice and KM pillars

Pillars	No of items	Mean (SD)			p-value
		Overall (n=331)	Academic (n= 164)	Non academic (n= 167)	
Learning	18	59.2 (12.7)	56.7 (12.1)	61.6(12.8)	0.000
Technology	10	37.1 (6.9)	36.2 (7.1)	38.1 (6.5)	0.009
Organization	11	34.9 (8.2)	32.7 (8.4)	37.0(7.5)	0.000
Leadership	14	43.9 (11.4)	41.3 (12.3)	46.6 (9.9)	0.000
KM –Total	53	175.2(33.4)	166.9(34.6)	183.3(30.2)	0.000

3.2 Ranking of the Existing and Desired pillars

Table 3.2 shows, the comparison of participants perceptions ranking of the four KM pillars from the most to the least (1 to 4) problematic in the current KM practices and the rank of the pillars participants desired the university should give future priority to improve KM practices in the university. The analysis computed by means rank and SD. The result of the study showed, from the overall study participants ranking of the pillars revealed, leadership was first then learning, organization and technology were ranked as second, third and fourth respectively. On the other hand, among the four pillars learning were ranked first for future improvement to improve KM practices in the university, then leadership, technology and organization were ranked from second to fourth respectively

Table 3.2 Rank of KM pillars based on the level of problems and priority

Pillars	Current level of problem				Future priority for improvement			
	Mean (SD)	Rank	X^2	p-value	Mean (SD)	Rank	X^2	p-value
Learning	2.4 (1.02)	2	238.78	.000	2.2 (1.0)	1	50.47	.000*
Technology	3.3 (0.97)	4			2.6 (1.3)	3		
Organization	2.6 (0.89)	3			2.9 (0.95)	4		
Leadership	1.8 (1.05)	1			2.3 (1.1)	2		

3.2.1 Comparison ranking of the existing KM Pillars

Table 3.3 showed the comparison of ranking between academic and non-academic participants in the four KM pillars by level of problem, from most problematic to least problematic (1 to 4). The academic staff ranked leadership to be the most problematic followed by organization, learning and technology. The non-academic staff ranked; leadership, learning, organization, and technology from the most problematic to the least (table 3.3).

Pillars	Comparison the current level of problem							
	Academic				Non-academic			
	Mean (SD)	Rank	X^2	p-value	Mean (SD)	Rank	X^2	p-value
Learning	2.7 (1.05)	3	116.14	.000	2.1 (0.93)	2	145.3	.000*
Technology	3.2 (1.07)	4			3.4 (0.87)	4		
Organization	2.4 (0.87)	2			2.8 (0.90)	3		
Leadership	1.7 (0.90)	1			1.9 (1.20)	1		

Table 3.3 Comparison ranking of pillars —academic vs. non-academic staffs

3.2.2 Comparison ranking of the desired Knowledge Management pillars

Table 3.4 showed the comparison of ranking between academic and non-academic participants in the four KM pillars (desired condition for future priority) to improve the current KM practices of the University from most to least (1 to 4). It was found out that ranking of the two groups was different. Academic staffs give credence to leadership in respect to KM to improve current practices and ranked leadership first; whereas, non-academic staff belief prioritizing technology in respect to KM will improve the current practices and ranked technology first.

Pillars	Comparison the Future priority							
	Academic				Non-academic			
	Mean (SD)	Rank	X^2	p-value	Mean (SD)	Rank	X^2	p-value
Learning	2.2 (1.0)	2	76.5	.000	2.3 (0.93)	2	43.3	.000*
Technology	3.1 (1.1)	4			2.2 (1.0)	1		
Organization	2.7 (1.0)	3			3.0 (1.0)	4		
Leadership	2.0 (1.1)	1			2.6 (1.1)	3		

Table 3.4 Comparison rankings of desired KM pillars academic vs. non-academic staffs

4. DISCUSSION

Examining the current KM practices in universities is important to identify the existing knowledge environment. According to Fathian et al (2008) to consider knowledge management assessment models, some points must be noted: the assessment models must be complete and be able to evolve the whole dimensions of organizational KM indicators. Also, the categories of the assessment model of KM must be separated from each other. By considering such points, the study utilized the pillars of the four KM pillars: learning, technology, organization and leadership; using three point score (3=neutral) in the five point Likert's scales as the cut-off point (average point) Moreover, from the simple ranking questions, the perceptions of staff in the current level of the problem and future priority for improvement among the four pillars was evaluated.

4.1 Evaluation of the Current KM Practices

The result of the present study revealed that the average score of the four pillars in relation to the current level of KM practices in the university was above average. This indicates that the respondents believed the level of KM in relation to these criteria was above average. Even though, the overall KM practices in the university were above average, the mean average score of academic and non-academic respondents to each pillar and the overall KM was different. The perceptions of non-academic staff to each pillar and KM in general was above average, whereas, the average score of the academic staff in relation to organization and leadership with respect to KM practices in the university was found to be below average with 32.7 and 41.7 respectively. This shows that the perception of academic staff to the total KM practices in Jimma University is below average. This is more or less comparable to the study done by Ejemeh & Gboge (2011) which assessed the level of application of KM principles and practices in Nigerian Universities using four assessment areas; Knowledge Awareness KM Tools, Knowledge Acquisition and Sharing and KM Audit).

Moreover, the result of the study shows there was a significant difference between academic and non-academic staffs in the perceptions of KM level and the four KM pillars. The results of another study done by Anvari *et al.* (2011) in Firoozabad Islamic Azad University in Iran in assessing KM practices in the university considering the four pillars (information, skills, culture and technology) showed significant difference in the perception of other staffs and lecturers.

Even though, the study need a more in-depth exploration for investigating the underlying causes of the difference in perception between the two group, the possible explanation for the variation is may be due to variation in understanding the KM principles, tools, techniques, approaches and challenges. Moreover, the variation might be due to the gap in understanding the advantage of KM approach to the university and the actual benefit. This may also indicate the difference between the two groups of respondents in the approach of KM practices in relation to learning, technology, organization and leadership. Thus, the respondents from academic staff, with better educational qualification and years of experience assessments result is more reliable because they might have more information and a deeper and wider understanding about KM practices and its long term benefits.

A comparative scoring, the perceptions of the two groups in the current level of the problem and future priorities among the four KM pillars is also different for the academic and non-academic staff of the university. The finding showed the non-academic perception of the existing problem and desired condition to improve the practices is controversial. On the other hand, in respect to academic staffs list of the existing problems and desired priority have similarity. Both group ranked leadership as the most problematic and technology as least problematic; But the two group vary in the ranking of learning and organization pillars. The finding of ranking the desired condition by academic staff is in line with a research done on working sector conducted by Calabrese (2005) to obtain each respondent's preferred rank ordering on the importance of the four pillars.

The analysis of the interviews showed that majority of the participants were aware about the need of sharing knowledge in order to accomplish and succeed in their work and to extend and sustained organizational development and competency. However, past best practices, research/project findings, were not adequately organized and documented so that they are easily located and accessed by the staff. Also the study showed lack of experience in information sharing on research findings and best practices across departments or colleges were an issue for not sharing the information and knowledge.

Moreover, the analysis of the interviews showed a clear difference with respect to KM enabler's perception. The non-academic staffs believe giving priority to technology with respect to KM is better to improve the current practices of KM in the university. Whereas, the academic staffs believe considering priority for leadership will bring better improvement in the current KM practices. Also, based on the analysis of the interview, awareness and approach to the knowledge resource, organizational culture, relay on technology and lack of trust and insufficient management support was among the major factors mentioned by participants that negatively influence the success of KM practices in the university. According to Yeh *et al* (2006), KM enablers should be clear in an organization, because they not only create knowledge but also prompt people to share their knowledge and experiences with others.

CONCLUSION AND RECOMMENDATION

From the result obtained, it is possible to conclude that the level of perception of KM practices between academic and non-academic staffs of Jimma University is not the same. The same holds true when it comes to ranking the perception of staff at the level of the current problem and desired condition to prioritize among the four pillars to improve future KM practices. According to the finding of the study, technology was least problematic and leadership the most problematic among the four pillar in respect to the current KM practices in the University.

Based on the findings, we recommendation the following to be considered in relation to the integration of the four pillars and KM practices to the university's practices, such as learning and teaching, research and innovation, community based education (CBE) and university management.

KM Learning

To develop the attribute of learning organization and organizational learning, the university need to give due consideration in identifying the knowledge gap and follow a systematic way to solve the gap. To cultivate the culture of learning in the university the following activities should be considered;

Creating a mechanism to connect people with people, through socialization like, training, workshop, facilitating across department research, making accessible of expert profile.

Creating a mechanism in connecting people with knowledge source. This can be applied through developing knowledge centre.

KM Technologies

The selection and providing of technology to staffs should be based on the capacity and task oriented. KM technologies that are appropriate for a particular KM system and environment and complex social systems and their impact on technology choice should be clearly identified and examined.

KM Organization process

Regarding the organization process, the value of knowledge creation and collaboration should be intertwined throughout the university. Since knowledge is everywhere, the university operational process must align with the KM processes.

KM Leadership

Considering the current problem the following suggestion can follow support for minimizing or solving the problem. Those are:
- Top management support and campaign for potential individual as well as organizational level learning
- Create mechanism in measuring and managing intellectual capital periodically (identified solutions by mapping existing tasks, procedures & processes.)
- Responsible body (Chief knowledge manager, knowledge editor, knowledge analyst etc.) should be appointed

REFERENCES

Akmar M.L. and Lee S.Y., (2004). Implication of Knowledge Management (KM) in Higher Learning Institution. [Available at]: http://www.educause.edu/ir/library/pdf/EQM0044.pdf

Anvari A., Abbas A. G., Moghimi R., Baktash L. and Mojahed M. (2011). An assessment of Knowledge Management (KM): A consideration of information, culture, skills and technology. *African Journal of Business Management,* 5(28), 11283-11294. http://www.academicjournals.org/AJBM [Accessed 16 November, 2011]

Calabrese, Francesco. (2000). "A suggested framework of key elements defining effective enterprise knowledge management programs." Doctoral dissertation, Engineering management and Systems Engineering, The George Washington University, Washington, D.C

David, De. L., (1997). Building the Knowledge-Based Organization: How Culture Drives Knowledge Behaviours, Working Paper, Centre for Business Innovation , Ernst & Young LLP.

David Skyrme Associates, (1999), "Measuring knowledge and intellectual capital," [Available at]: http://www.skyrme.com.

Ejemeh, D.K. and Solomon Gboge, S.K., (2011). Towards sustainable Development: an assessment of knowledge management initiatives in Nigerian universities. *Journal of Sustainable Development in Africa,* (13(3), ISSN: 1520-5509.

Lubega T. Jude, Omona Walter and Van der Weide, Theo, (2011). Knowledge Management Technologies and Higher Education Processes: Approach to Integration for Performance Improvement. *International Journal of Computing and ICT Research*, Vol. 5 Special Issue, 55-68.

Mikulecka, J., & Mikulecky, P. (2000, September). University knowledge management-issues and prospects. In Djamel A. Zighed, Jan Komoroswki, Jan Zykow, Principles of Data Mining and Knowledge Discovery 4th European Conference Proceedings, PKDD (pp. 157-165).

Pircher, R. and Pausits, A., (2011). Information and Knowledge Management at Higher Education Institutions. *Management Information Systems,* 6(2), 008-016.

Park, H., (2005). Knowledge management technology and organizational culture" in M.

Stankowsky (ed.), Creating the discipline of knowledge management: the latest in university research, Amsterdam , Elsevier Butterworth-Heinemann.

Senge, P. M. (1990). The Fifth Discipline: The Art and Practice of the Learning Organization. New York, .Currency Doubleday.

Stankosky, D. Sc., (2005). Advances in Knowledge Management: University Research Toward an Academic discipline. In Stankosky, M. (Ed.) Creating the Discipline of Knowledge Management. Washington, Elsevier Butterworth-Heinemann.

Sekaran, U., (2000). Research Methods for Business: A Skill – building approach. New York: John Wiley and Sons. Inc., 308-313.

Theriou, N., Maditinos, D., & Theriou, G. (2010, January). Knowledge Management Enabler Factors and Firm Performance: *European Research Studies*, 14(2), 1-20).

Yang, Tai Ning, Chang Hsiao Chen, Lin Shou Yen and Tsao Chiao Lun, (2011). Knowledge Creation and Intellectual Capital on Securities Investment Services, *African Journal of Business Management,* 5(3), 924-933.

Yeh, Y., S. Lai, and C. Ho, (2006) "Knowledge management enablers: a case study", *Industrial Management & Data Systems*, 106(6), 793-810.

CHAPTER 10:

Knowledge, Attitude and Utilization of Information Communication Technologies (ICTs) in an Ethiopian Medical Teaching Hospital

Melisachew Adane, Lemma Lessa and Solomon Shiferaw

1.
INTRODUCTION

Developments in information and communication technology occur at an astonishing rate. This has had huge implications for medical practice throughout the world (Samuel et al, 2004). Medicine has always been an "information-intensive" occupation, and the penetration of information technology into practice and education is generally welcomed. The advances in technology provide greater ease of access and use to exploit the benefits of computing for medical education as well as quality health service delivery purposes (Kenneth & Michael, 2000; Samuel, et al, 2004). This trend of increased use of computers in medical practice makes it important for medical school graduates to develop and enhance computer skills for their future practices (Amy, et al, 1997).

Although the importance of computer literacy in today's rapidly changing environment cannot be denied, institutional provision of opportunities and inclusion of such skills into their curriculum for medical students to acquire the necessary technical skills has been comparatively slow and inconsistent (Kenneth & Michael, 2000; Thomas & Richard, 2006). One of the basic concerns with this regard is that the spread of information and communication technologies in developed countries is leaving the rest of the world behind. In most of the developing countries including Ethiopia many programs have concentrated on increasing the number and spread of ICT infrastructures without adequate effort on the capacity building. These may create a gap in ability and utilization of ICT and also reduces the impact of Information technology use, measured by financial, economic and clinical returns. In other words, equipment alone is useless unless people are able to use it effectively and informed of the potential benefits of its use (Samuel et al, 2004).

The Government of Ethiopia believes that exploiting information technology is central to promote growth and reduce poverty. One of the ICT strategies in Plan for Accelerated and Sustained Development to End Poverty (PASDEP) is mainstreaming the use of ICT in all sectors of the economy. A central part of this strategy is applying the principle of modern ICT to the delivery of services and administration of government, to improve effectiveness and reduce costs.

Accordingly, health is one of the main service delivery sectors that need mainstreaming the ICT to provide quality health care delivery. Equipping the health professionals with appropriate knowledge and skill of ICT during their stay in the medical school is vital to implement the information technology in the health sector as to the national strategy (Ministry of Finance and Economic Development (MoFED) of Ethiopia, 2006). However, the reality in the ground is different according to the ICT penetration and usage base line study conducted in Ethiopia that indicated only 2% of the sampled health professionals have formal college or university ICT training.

This may be an evidence to say much should be done in the medical institutions in the area of increasing ICT skills for the prospect graduates as well as the existing health care providers. The skill and utilization patterns of ICT among medical students is not known and to my knowledge there is no published reports on the knowledge and utilization patterns of ICT among medical students in Ethiopia particular in Black Lion Hospital. Hence, this assessment will identify factors contributing to the poor ICT knowledge, skill and utilization in the medical set up (Mulat & Tadesse, 2002).

It is well known that computers influence every sphere of human activity and bring many changes in medical education, healthcare and scientific research etc. Computers can perform a wide range of activities that save time and help health care providers to be engaged in other patient care activities. The availability of quality applications for medical education in both the basic and clinical sciences makes it feasible for an institution to incorporate such applications into the existing curriculum (Amy, et al, 1997; Emans, et al, 2004). Accordingly, it is necessary to ensure the knowledge and utilization of ICT among those who deliver the service. Therefore this study is going to assess the knowledge of medical students on ICT and their utilization status and also the possible factors that influence using ICT for their academic purpose and service delivery needs. In line with this, the findings will also provide baseline information for medical faculty and health facility administrators in planning ongoing computer training for medical students as well as health care providers currently working in the health facilities.

2.
LITERATURE REVIEW

The emerging need of computer knowledge has made an impact in every field, including the medical world (Jha, et al, 2004). Since the development of the computer and the evolution of the Internet, Information Technology (IT) has had a positive impact on health care delivery systems worldwide, particularly in the areas of disease control, diagnosis, patient management and teaching (Ibrahim, et al, 2004). In general, clinical practice has been tremendously improved by the technological interventions (Ime, et al, n.d.). As a result of this revolution the application of ICT especially in the areas of information access, storage, retrieval, analysis and dissemination of information is becoming a routine activity in the health care system. This makes it almost mandatory for the healthcare professionals to be well versed with this technology in the developed world (Jha, et al, 2004).

Moreover computerization could resolve certain problems and derive benefits including reduction in clerical work required of professionals, reduction in printed forms, centralized patient care data etc. (Emans, et al, 2004). The development of online databases plays a vital role in packing and delivery of medical research as the same time it allows medical professionals throughout the developed world immediate access to hundreds of e-journals at the touch of a button, a striking contrast to the plight of many of their colleagues in developing countries who are forced to search empty libraries (Samuel, et al, 2004; Nkeiruka, et al, 2008).

Rapid advancement in information technology and the increasing availability of electronic health information are revolutionizing health care systems worldwide. Innovations in technology have made portable electronic devices, network database applications, electronic medical records, and computer software programs increasingly integrated into many health care settings. These changes create the educational need for health care providers to become proficient at appropriately using technology to deliver high quality health care services. However, the impact of ICT on medical practice in the developing world is not significant due to lack of awareness and access to computer (Tina & Scott, 2007).

Accessibility of ICT

The use of computer and Internet technology by health science students will result in more effective medical education, including teaching, medical examination, and diagnosis of disease. However, these gains will only occur when students have increased access to these technologies (Grace, 2003). One of the central factors identified for the skill of computer among physicians is the ownership of a personal computer. This is because it is associated with better information-handling knowledge and favourable perception of the computer-based record system (Ime, et al, n.d.).

Several studies in the accessibilities of IT among medical students and health workers showed a magnificent difference between the developed and developing world. In a survey of physicians across eleven North American, European and Asian countries in 1998, 80% of physicians were found to own a computer and 44% of these physicians had accessed the Internet and their predominant place of Internet access was in the home. In another survey of European Universities Skill in ICT of students and Staff (SEUSISS) project (2001-2002) the number of personal computer (PC) ownership at the start of studies varied from 54% (Abo, Finland) to 89% (Groningen, Netherlands). Recent studies have however shown remarkable improvements in these figures. But much is not reported in the literature about the level of Internet access amongst doctors and the use of electronic medical records of health facilities in Africa (Jens, 2004).

A study conducted in Medical university of Vienna showed almost all students (94%) have access to a privately owned PC, which is either owned by the students themselves (74%) or shared with family members or roommates (20%). Only 5% rely primarily on public computer facilities. In the same study the great majority of students also have access to the Internet (Thomas & Richard, 2006). Similarly a longitudinal study in Aarhus, Denmark showed among the total first year medical students, 71.7% indicate they had access to computer at home. In this study Internet access at home was also assessed and it was found to be increased from 20.4 to 62.9% in the study period and there was an even more pronounced increase in the use from any location, of Internet and e-mail (Jens, 2004). On the other hand in developing countries, the Internet is still only available to a minority of health professionals, and often it is not available at the point of care (Grace, 2006).

The access to computer of medical students in Africa is lagging behind when compared to the developed world. A study in the MUCHS, Tanzania showed the medical student to computer ratio is 100:1. It is too far behind when it is compared to 35:1 in Portugal, 9:1 in the UK and 5:1 in Norway. With regard to ownership of computer at home 76% of MUCHS students didn't have a computer at home. This figure is in stark contrast with the availability of computer at home which is 71.7% Aarhus, Denmark and 86% in California, USA (Jens, 2004); Samuel, et al, 2004). Similarly a study in Ile Ife university teaching hospital, Nigeria showed only 26% of students owned a computer (Ibrahim, et al, 2004). Even if computer access is vital in acquiring skill of ICT, almost all of the studies conducted in Africa indicated that access to computer is very limited.

A base line study on ICT penetration and usage in Ethiopia indicated that 51% of health workers gain access to computers in their work place. Private computer centres are the second most important place (46%) and only 12% of the sampled health professionals had computer access at home of which majority of them are from Addis Ababa. It should also be realized that access to printed materials is very much limited in the health facilities, especially outside of Addis Ababa. As information is critical in the operation of health professionals every effort should be made to improve access to computers and the Internet at the work place. Thus Health workers will constantly update themselves about new developments (in the area of treatment and nature of various diseases) with the help of computers and Internet (Mulat & Tadesse, 2002).

Therefore providing students with computer access in addition to the theoretical training by placing computer stations in the library or by developing a dedicated computer laboratory space have been the most common solutions adopted by institutions to ensure the skill of medical students on ICT before they leave the campus. As a result of the aforementioned facts recently some medical schools have considered policies recommending that students acquire a PC for their medical education, or even requiring them to do so (Amy, et al, 1997).

Knowledge and Attitude towards ICT

In this ICT era the need for medical students to be computer literate is no longer an issue for debate. Currently using computerized medical records, retrieving patient data at a distant and accessing medical journals and literature electronically is common. Hence, acquiring basic knowledge on computer and understanding the basics of Internet among physicians is crucial (Brenda, et al, 2002). However, several studies in the literature suggest many medical students feel that they lack computer skills and majority of them are interested in learning more about computers while attending medical school (Sue, 1999).

A study conducted in Punjab, India showed that majority (75%) of staff nurses had good computer knowledge and 21% had average computer knowledge in clinical care setting. All the nurses had positive attitude towards computers (Emans, et al, 2004). Another study from Chennal, India indicated that 42% of Undergraduate (UG) and 47% of Postgraduate (PG) medical students felt that working with computer gave them a lot of self-confidence. Concerning Knowledge 40% of all medical students did not have any knowledge of database and spreadsheet (Jha, et al, 2004). Similarly the longitudinal study conducted in Aarhus, Denmark indicated that 86% of male and 76% of female students had positive attitude towards use of ICT resources as a supplement for their medical education (Jens, 2004).

Contrary to this a study from Nigeria teaching hospital showed that from the total respondents only 19% and 40% of the medical students demonstrated a good knowledge on computers and had positive attitude towards IT respectively (Ibrahim, et al, 2004). The study from MUCHS, Tanzania indicated that only 52% of students felt that they understood the basic terminology and concepts of computing (Samuel, et al, 2004). Similarly a study from a resource poor setting showed that 60% of the students who were familiar with computers acquired their knowledge through self-learning efforts while 37.5% attended a formal training.

Based on the ICT penetration base line survey conducted in Ethiopia the overall computer literacy among health professionals was 39% with no visible differences between Addis Ababa and regional towns. Personal effort is the main method (67%) of acquiring ICT skill, which is similar to other studies followed by short-term computer training (33%), formal College or university ICT training is reported by only 2% of the respondents (Mulat &

Tadesse, 2002; Nkeiruka, et al, 2008). These all facts in the literature indicate delivering computer courses at the medical institutions is neglected. Moreover majority of the literatures suggest the need for training of physicians in the use of computer in the medical education. To that end, determining the gap of medical students for acquiring computer skill in terms of training or accessibility should be assessed before starting any intervention (Amy, et al, 1997).

Utilization of ICT

Clearly, medical students need to acquire computer and information management skills at the beginning of their medical education. As indicated in many literatures most of the latest reference materials are accessible electronically; this is also an assertion supported by a recent two-year survey at the University Of Illinois College OfMedicine at Rockford (Kenneth & Michael, 2000). In support of the above fact a study conducted in Austria in 2004 showed that 75% of university and high school students used a computer daily for different purpose especially for e-mail communication (94%), Internet for information research (97%) and use of word process is very common (82%), but students are less familiar with other program types. (Thomas & Richard, 2006). Majority of the studies conducted in the developed world showed the skill of students in using ICT is high which were showed by 84% of undergraduate students in Glasgow, UK, 95% undergraduate dental students in Oulu, Finland. There is also an encouraging trend of ICT utilization in some of the East and Central Asia countries like 94% and 95% of medical students in Malaysia and Saudi Arabia respectively use computers for their medical education (Ibrahim, 2002; Grace, 2003).

In spite of the limited studies conducted in the developing world especially in Africa, some of the studies showed that the skill and utilization of ICT among medical students as well as health care providers are very poor. In support of the above idea a study from Nigeria teaching hospital on Computer and Internet use by first year clinical and nursing students showed that, only 43% of students could use the computer. Odusanya and Bamgbala in the same institution found that 80% of the medical and dental students in their final year had used the computer; however, the use of software applications was poor, with computer games being the most frequently used (19%) followed by word processing software (18%). The Internet and email were used by 58%, but only 23% had used the Internet for medical research (Grace, 2003). In similar context a study from Tanzania medical university college indicated that about 74% of medical students never use a computer as part of any course either at school or university. Out of those who are using (25%) the median hours per week of computer use was 3.8 (Kenneth & Michael, 2000; Ibrahim, et al, 2004). At this time their computer skill was also measured and found to be very low. The method of measuring skill is adapted from the Centre for Health Informatics and Multi professional Education (CHIME) in UK, students with an overall score of less than 10 are considered to have low skills and are offered peer mentoring training. Using this criterion, around 50% of the Tanzanian medical students would fall into the low skills category compared with 9% of first year UCL medical students in 2002. All these studies concluded that utilization of ICT was poor amongst the resource poor sub Saharan African countries (Ibrahim, et al, 2004; Samuel, et al, 2004).

Similarly in Ethiopia only 33% of the health workers use computers for various purposes. The single most important purpose of using computers is word processing and related activities (office tools) for 81% of the reporting respondents. Some 26% of the sample respondents use the Internet. As expected, e-mail is the most important reason for using the Internet (90%). The barriers identified for ICT utilization in the health sector during the base line survey on ICT penetration and usage in Ethiopia were Lack of necessary equipment(accessories) and absence of ICT strategic plan. In addressing the possible factors for ICT utilization the current ICT strategy of the country is encouraging especially in scale up of ICT infrastructures at all level of service delivering sectors. Hence the medical institutions should take this as a good opportunity to incorporate computer courses in their medical education then medical students may acquire appropriate skill for their future carrier (Mulat & Tadesse, 2002; Ministry of Finance and Economic Development (MoFED) of Ethiopia,2006).

3.
METHODOLOGY

The study was cross-sectional survey by design complimented with qualitative in-depth interview. The study was conducted at Black Lion Specialized Hospital in Addis Ababa. All students and health care providers in the Faculty of Medicine including residents, medical doctors and nurses, either attending their medical education in the faculty or providing health care service in Black Lion Specialized Hospital during the study period, were the source population. Among all the students in the Medical Faculty, Intern and clinical year-1 medical doctors, final year nursing students (Both clinical and midwives) and residents attending their education and health care providers working in Black Lion Specialized Hospital were selected for the study. These groups were selected purposively mainly to increase the likelihood of having more respondents with computer exposure. Then the total sample size was proportionally allocated based on the number of students and health care providers in each category. Finally the respondents were selected randomly from each category.

Both quantitative and qualitative techniques of data collection were employed in the study. Structured questionnaire encompassing all the variables of interest were adapted from other related articles for its consistency reliability next to standard questionnaire and it's modified to the context to fit the study population. Semi-structured questionnaire were developed to guide the qualitative data collection. In-depth interview was carried out with department heads, Instructors and students of the Medical Faculty and physicians working in Black Lion Specialized hospital to gather qualitative information. The in-depth interview was conducted using separate interview guide for each participant from each category of respondents by the principal investigator with the assistance of supervisors. Tape recorder was not used because of the refusal of the key informants to be recorded

Quantitative data were initially entered and cleaned using EPI-6 DOS version and exported to SPSS 15.0 for analysis. Frequency tables, proportions and crosstabs were used for the descriptive analysis. And for presentation tables and different type of graphs were employed. Association between independent variables such as, computer training, computer possession, having computer course, place of high school completed, category of study and socio-demographic characteristics of students and outcome variables of Knowledge, Attitude and Utilization, was examined using odds ratio, X^2 test and logistic regression when it is appropriate. For multivariate analysis the necessary adjustment was done for the possible confounding factors to identify the predicting factor for knowledge, Attitude and utilization of ICT among students category. Hence, internal comparison was done based on the adjusted odds ratio. In line with this the significance level was set at p-value = 0.05.

4.
RESULT AND DISCUSSION

In Ethiopia there is no as such study conducted assessing the knowledge, attitude and utilization of ICT among medical students. This study contributes as base line information on the knowledge, attitude and utilization of ICT among Medical students and health care providers. The study indicated that 24.9% of the respondents had satisfactory knowledge and majority of them were undergraduate and residents. Of which almost half of them 41.1% were interns and clinical year medical doctors and this may be due to their exposure from their high school stay. This figure is comparable with the finding from Nigeria teaching hospital where 18.9% of medical students and health care providers had good knowledge (Emans, et al, 2004). However, this figure is less than the data obtained from the ICT penetration survey among HCPs in Addis Ababa and four towns of regions which was about 39% (Kenneth & Michael, 2000). Similarly, a study conducted in Nigeria was also greater than this finding 50.6% of clinical year medical students (Nkeiruka, et al, 2008).

In this study 56.7% of the respondents know basic ICT terminologies which are comparable with the study done in Tanzania College of Health Science, 52% of 4th year clinical year medical students felt that they understood the basic terminologies and concept of computer (Samuel, et al, 2004). The study also showed that younger age groups are relatively better knowledge of ICT than the older one. The reason behind may be younger ages are more favourite to new technology than the older ages. In support of this finding, a study conducted in Nigerian teaching hospital showed that younger respondents tended to have multiple accesses to internet than older age (Ime, et al, n.d.). The study also indicated that males from gender category are superior in knowledge level than females. The result is consistent with other studies conducted in Nigeria and Yemen (Abdula, 2008).

The study revealed that 62.2% of respondents had different type of formal and informal ICT trainings; however, training status was not translated to better knowledge among the respondents who have different type of training, only 26.5% of them score above the 3rd quartile. Similarly, in a study conducted at Ile Ife university teaching hospital, only 26% of medical students and 27% of physicians had good knowledge after having different type of training (Ibrahim, et al, 2004). A study from Punjab, India also indicated that type of computer training received by the nurses does not influence the nurses overall computer knowledge (Emans, et al, 2004). The possible reasons may be the type of training most of them had informal training less than 6 month and the time they received; majority of them took the training during their high school stay. On top of this, quality of the training, content of the training, mode of delivery, instructors' capacity, allocated time for theory and practical session during the training and gap between the training time and the actual use of computer are also contributing factors for knowledge status among those who had short term training.

This study showed that 67% of students had computer course in their current study. On the other hand their knowledge had no significant difference with those who had no computer course. These may be explained by the adequateness of the course and also the time the course is delivered. About 89.1% of medical students who took the course were not satisfied with the computer lab session. The in-depth interview also support this in which it is explained that the course delivered was simply introduction to computer in two and three credit hours and the allocated time for computer lab session was only two weeks.

In this study 51.7% of the total study subjects had at least one personal computer in their home. This figure is by far better than the result from ICT penetration in Ethiopia, computer access at home was reported by only 12% of the total respondents of HCP this may be the cost of the computer in previous times and also most of the respondents in the current study were final year students especially residents who possess the larger proportion of personal computer (Mulat & Tadesse, 2002). A survey of trainees in a Nigerian teaching hospital indicated the same figure, 51.7% of participants had personal computer (Ime, et al, n.d.). Similarly, the data from Aarhus, Denmark showed that 71.7% of first year medical students had access to computer at home (Nkeiruka, et al, 2008). In contrast to this finding, a study in Tanzania showed that about 76% of the medical students having no computer at home (Samuel, et al, 2004).

The current study showed that possession of personal computer is one of the determining factors to have better score of basic ICT knowledge. In support of this study the finding from Nigerian teaching hospital indicated that ownership of personal computer is statistically significant association with knowledge of computer and longer duration of practice (Ime, et al, n.d.).

Access to computer in the Medical Faculty is very limited as disclosed in this study. The in-depth interview pointed out that no computer is dedicated for academic purpose in each department except in the computer lab. This explanation is similar with the study result from ICT penetration survey in Health sector of Ethiopia where it is stated as "accesses to the available computers in the health facilities were often limited to secretaries of facility officials" (Mulat & Tadesse, 2002).

Internet access was found to be 74.8% of the study subjects. The most frequently mentioned place for Internet access was Internet café (46.6%). This finding is much better than the result from ICT penetration base line survey in Ethiopia where it is showed that only 26% had access to Internet. However, the place for Internet access is comparable with the current study in that around 48% were accessed it at Internet café (Mulat & Tadesse,

2002). In terms of access, this study is also better than a study conducted in Nigeria among first year clinical and nursing students in which around 60.7% had access to Internet and the common place of access is cyber café (87%) (Grace, 2003).

In this study 83.7% of all respondents had positive attitude towards Information Communication Technology. This finding is more or less comparable with the study conducted in Ludhiana, Punjab, India in which all the nurses (100%) had positive attitude towards computers in health care setting hospital (Emans, et al, 2004). Similarly, a study done in Athrus, Denmark showed that majority of the first year medical students (68.4%) would like to replace traditional teaching with use of computers if possible and 88.1% of dental students in Egypt had also positive attitude towards computer (Nkeiruka, et al, 2008; Tanawi & Saleh, 2008). Therefore, we can say that even if there is limitation in access to computer most of the medical students are willing to know about ICT facilities and also to use it for their academic purpose and health care delivery needs.

The highest level of competence reported by the study subjects were word processing 44.3%, email use 53.9% and Internet browsing 45.6%. Which was average and above competence of application skill. This finding is comparable with the study conducted in Nigeria and Tanzania word processing competence 60%, e-mail 75% and Internet 58% (Samuel, et al, 2004; Nkeiruka, et al, 2008). In addition, graphics and MS-PowerPoint competence is higher among residents because of the fact that their field requires a lot of presentations and seminars.

Regarding the way of acquiring skill, 67.3% of Internet browsing, 54.6% of graphics and MS-PowerPoint presentation and 45.7% of word processing skill was acquired by personal effort. This result may be explained by most of the students use the word processing and MS-PowerPoint for report writing and preparing presentation and regard to internet majority of them use email. The study has similar result with the ICT penetration survey in Ethiopia, personal effort is the main method (67%) of acquiring ICT skill followed by short-term computer training (33%) and short-term in-house training (21%) among health care providers in Addis Ababa and four main regional towns (Mulat & Tadesse, 2002). Comparably studies in Nigeria and Tanzania indicated that self-effort contributes around 60% of the ways of acquiring skill and 68.9% in southern US medical university (Deiride, et al, 2000; Samuel, et al, 2004; Nkeiruka et al, 2008).

In this study 32.6% of study subjects were categorized under good utilization rate. Majority of the users were residents 52.3%. This is explained by the access level of computer among the respondents. Most of the residents have probably better income compared to the undergraduate students. This also reflected by their computer possession 93.3% of all residents had their own computer at home. Similarly, 33% of the health workers use computers for various purposes mainly for word processing and related activities (office tools) in ICT penetration survey of Ethiopia and in Zaria, Nigeria 26.7% of medical students utilize computer of them 60% comfortable in word processing. In contrary to this finding the study conducted in Chennai, India 94% of both undergraduate and post graduate students were found to utilize the computer for desktop usage. The Postgraduate medical students were found to use the computer more frequently as compared to undergraduate medical students (Mulat & Tadesse, 2002); Jha, et al, 2004); Nkeiruka, et al, 2008).

This study indicated that Internet is used by 87% of respondents and 80.4% of the study participants had email address for their mail communication. The email was the most common application used (90.1%) followed by research and education (66.1%) in this study samples. Similarly, a study conducted in public medical school of the southern United States in 2000 showed majority of students had used email (97%), educational software (75%) and conducted literature search (88%) (20). The study also concordant with the result from Nigeria where 76.4% of first year clinical and nursing students in Ibadan (Grace, 2003) and 75% fourth year medical students in Dare-Selam, Tanzania (Samuel, et al, 2004) have used email. The explanation for poor Internet use among students for academic purpose may be due to the fact that Internet connection is poor, slow and high cost in the Internet café. This makes difficult to download literatures and other documents from the Internet. As a result, most of the students inclined to use hardcopy for their education as a text or reference. Regarding the email service it is possible to be served with low level Internet connection in the Internet cafes.

The study also pointed out that computer possession is one of the main factors that resulted in better utilization status of students. Accordingly, 53.6% of respondents of those who have personal computer had better ICT utilization status. A study in Nigeria and New Jersy, Startford university of medicine showed that ownership of personal computer is highly associated with longer duration of practice (Lloy & Sherry, 2006). There are various reasons for not using computers and the Internet. Of which lack of access to ICT facilities, inadequate training and high cost are among the major constraining factors. (Mulat & Tadesse, 2002; Abdula, 2008).

In this study it is indicated that knowledge of respondents are one of the driving factor of utilization of ICT among students in the academic as well as in the health service delivery needs. Among respondents who had good knowledge 52% of them use computer satisfactorily. Likewise a study conducted in Helsinki, Finland demonstrated that basic computer skill was highly associated with utilization of ICT facilities (Kalle & Matti, 2006).

From the in-depth interview it is indicated that majority of the key informants explained that, majority of students in the medical faculty needs to have competence of computer skill like statistical analysis software and online literature searching. In consistence with this finding a study in Hadramout university of Yemen showed that 74.3% of medical students need to be provided with training on online literature searching (Abdula, 2008).

4.
CONCLUSION

To design proper interventions of improving the awareness and utilization of ICT in the medical institutions and health care delivery system, it is worth to assess the knowledge, attitude and utilization of medical students and health care providers. Such kind of empirical investigation can be a spring-board for the appropriate interventions like curriculum revision or in-service training.

The results of this study have important implications for the knowledge and utilization status of health professionals in the medical institutions and health facilities. The study indicated that majority of the medical students and health care providers had low level of basic ICT knowledge and only half of them are familiar with the basic ICT terminologies. Majority of the respondents have received formal or informal ICT training, however their training did not differentiate them from those who have no training in the basic knowledge of ICT. Almost all of the study subjects have positive attitude towards ICT and willing to have the appropriate skill for their academic purpose as well as for their future carrier. Ownership of personal computer at home was found to be one of the determining factors to have better knowledge of computer. Almost all of the undergraduate students who received the ICT course were not satisfied with the course and uncomfortable with their computer lab sessions. Possession of personal computer was relatively better among residents which helped them to have relatively better utilization rate compared with the other groups.

The younger age groups had more likely to have better knowledge attitude and utilization of ICT applications compared to older ages. The study showed that access to computer and other ICT facilities is positively associated with participants' knowledge and utilization of computer. However, access to ICT facilities in the medical faculty was very limited and there is no computer dedicated for academic purpose in each department. Internet access was limited at Medical Faculty and majority of the respondents' access in the Internet café mainly for email service. The highest level of competence, reported by the study subjects were word processing and Internet browsing for email service and most of them acquired it through personal effort. Computer possession is one of the main factors that resulted in better utilization status of respondents. In this study knowledge on basic ICT is found to be one of the contributing factors for using the existing ICT facilities among undergraduate and postgraduate students in AAU, Medical faculty. There are various reasons for poor utilization of computers and the Internet. Of which lack of access, inadequate training of ICT and high cost appeared to be the major constraining factors.

REFERENCES

AAU, MF. The Addis Ababa University Medical Faculty Black Lion Specialized Hospital (2008) A 34 years journey /from 1973-2008/ AAU, MF. Addis Ababa.

Abdula, S., (2008). Using computer and internet for medical literature searching among medical students in Hadramout university, Yemen. *Online J Health Allied Scs.*;7(1), 6 URL: http://WWW.ojhas.org/issue25/2008-1-6.htm

Amy, V., Carol, L., Robert, R. & Lois, M., (1997). Implementing a Requirement for Computer Ownership: One Medical School's Experience. Med Educ Online [serial online]; 2, 4: 1-8; URL: http://www.utmb.edu/meo/

Brenda, L., Jeanne, B. & Carol, L., (2002). Using a decade of data on medical student computer literacy for strategic planning. *J Med Libr Assoc*, 90(2), 202-209.

Deiride, C., Theodare, W., David, A. & Tack, E., (2000). Variables that may enhance medical students' perceived preparedness for computer based testing. *J AM Med Inform Assoc.;* 7(5), 469-474.

Emans, E., Rajinder, M. & Veena, B. (2004) An Exploratory Study to Assess the Computer Knowledge, Attitude and Skill among Nurses in Health care Setting of a Selected Hospital, Ludhiana, Punjab, India: *MEDINFO:* 107(2), 1304-1307.

Grace, A., (2003). Computer and Internet use by first year clinical and nursing students in a Nigerian teaching hospital. *BMC Medical Informatics and Decision Making*, 3(10), 1-7.

Grace, A., (2006). Use of the Internet for health information by physicians for patient care in a teaching hospital in Ibadan, Nigeria. *Biomedical Digital Libraries,* 3(12), 1-9.

Ibrahim, S., Ikechi, T., Emmanuel, A., Fatiu, A., Abubakr, A. & Adewale, A., (2004). Knowledge and Utilization of Information Technology Among Health Care Professionals and Students in Ile-Ife, Nigeria: A Case Study of a University Teaching Hospital. *J Med Internet Res;* 6(4), e45.

Ime, E., Oluwakayode, O., Oladimeji, F. & Sesan, M. (2008) Computer Use Among Doctors In Africa: Survey Of Trainees In A Nigerian Teaching Hospital. *Journal of Health Informatics in Developing Countries,* 2(1), 10-14.

Ibrahim, M., (2002). Computer skill among medical learners: A survey at King Abdul Aziz University, Jeddah. J Ayub medc coll. 14(3):13-15; URL: http://www.ayubmededu.pk/JAMC/PAST/14-3/IrahimMansoor.htm

Jens, D., (2004). Experiences and attitude towards information Technology among First year medical students in Denmark: Longitudinal questionnaire survey. *J Med Internet Res.,* 6(1), e10.

Jha, P., Paul, L., Ojha, P., Sen, S.& Sinha, N. (2004) The use of computer by medical students in Chennai, India. Medindia.com, Network for Health. [Available at]: http://www.medindia.net/articles/computer_using_medical_students.asp

Kalle, R. & Matti, A., (2006). A survey of the use of electronic scientific information resources among medical and dental students. *BMC Medical Education,* 6(28) URL: http://WWW.biomedcentral.com/1472-6920/6/28

Kenneth, E. & Michael, S. A., (2000). Two-year experience teaching computer literacy to first-year medical students using skill-based cohorts. *Bulletin of the Medical Library Association,* 88(2), 157-164

Lloy, J. & Sherry, C., (2006). Computer assisted instruction: a survey on the attitudes of osteopath medical students. *J AM Osteopath Assoc.,* 106(9), 487-494.

Mulat, D. & Tadesse, B., (2002). ICT Penetration And Usage In Ethiopia: Baseline Study. Addis Ababa: Unpublished.

Ministry of Finance and Economic Development (MoFED) of Ethiopia (2006) A Plan for Accelerated and Sustained Development to End Poverty (PASDEP). Addis Ababa: MoFED, Volume I.

Nkeiruka, A., Kene, T. S. & Emmanuel, A., (2008). Computer knowledge amongst clinical year medical students in a resource poor setting. *African Health Sciences,* 8(1), 40-43.

Sue, H., (1999). Assessing and enhancing medical students' computer skills: a two-year experience. *Bull Med Libr Assoc* , 87(1), 67-73.

Samuel, M., Rob, M., John, C., Jaime, M., Eoin, J.W. & Pejman, A., (2004). Assessing computer skills in Tanzanian medical students: an elective experience. BMC Public health, 4(37), 1-3.

Tanawi, E. & Saleh, S., (2008). Attitudes of dental students towards using computer in education-a mixed design study. WHO, *Eastern and Mediterranean Health Journal,* 14(3) URL: http://WWW.emro.who.int/useinternet/

Thomas, M. & Richard, M., (2006). Computer literacy and attitudes towards e-learning among first year medical students. BMC Medical Education, 6(34), 1-8.

Tina, P. & Scott, R., (2007). Instructional designs and assessment An Interdisciplinary Online Course In Health Care Informatics. *American Journal Of Pharmaceutical Education;* 71(3), Article 43.

CHAPTER II:

Multi-criteria decision modelling for infrastructure development: A case of Ethiopian highway rehabilitation projects

Tamirat Fikre Nebiyu

INTRODUCTION

Ethiopia is a second populous developing country in Sub-Saharan Africa. Arguably, it is experiencing fast economic growth which is also accompanied with a huge demand for transport infrastructure. Despite the undergoing ambitious railway development, the sector of road transport remains the nation's single most important means for travel and transport; and it accounts for more than 95% of the country's domestic passenger and cargo traffic (Authority 1998; Worku 2011). However, the average road density was only about 38.6 km per 1000 km² in 2007(Shiferawa et al. 2012) which is still less than the average of 50 km per 1000 km² for sub Saharan Africa. Moreover, Ethiopia currently has only two express ways.

These are: Addis Ababa ring road and Addis - Adama express way which is now under construction. The regional highways have only two lanes; one in the opposite direction of the other. The share of improved national road length was grown: from 31% in 1995 to 40% in 2007 (Shiferawa et al. 2012) and this value was projected to be 50% in 2012. Majority of the highways, however, are still either in poor condition or under construction. All these evidences indicate that the country is struggling to attain the high necessity of improving the quality of road infrastructure.

Efficient use of available finance is crucial in road construction industry because resources are scarce. Road quality improvement[2] is one of the ultimate goals of Ethiopian Roads Authority- ERA. In Ethiopia, a number of highways need to be either upgraded or rehabilitated[3] . A report of road development plan performance in ten years: 1997-2007 (Africa and Authority ; Worku 2011) informs us two major challenges (Shiferawa et al. 2012). At the *first* place the number of highways to be rehabilitated has been increasing. And *secondly,* those highways which deserve urgent improvement must be prioritized within themselves because they are many. Both cases are challenges to Ethiopian road sector development program - RSDP. Let's explain them in a clearer manner.

Challenge 1 In Ethiopia several highways need improvement: As the country's growth becomes swift, ERA has become busy with building several new regional highways while the quality of existing once reduces through time. Ethiopian highways need continuous improvement because roads do wear out fast. This is mainly because of the low quality of construction: Evidently, more than 55% of federal road are not in a good quality, more than 85% of the total road surface is unpaved and there exists unbalanced axial load which has a damaging role to the highways(Authority 1998). In addition to that, the capacity of the government for rehabilitating all the roads is pretty low. For instance, from 1997 to 2007, the share of expenditure on highways from foreign financial sources was 52.56% (Worku 2011). This implies that the county's financial capacity is very scarce to rehabilitate or generally to improve highways. Therefore, for effective use of both internal and "unpredictably" external finance, Ethiopia has to prioritize highway rehabilitation projects.

Challenge 2 Ethiopia lacks a decision support tool for prioritization: Ethiopia has never had a well reputable reference document on road transport planning until 2006. The first comprehensive planning manual which was prepared and published by Ethiopian Roads Authority – ERA was launched in May 2006, (Becker and Demissie 2006). This implies that the

---

2 Road quality improvement means: construction, upgrading, rehabilitation, or maintenance
3 New road construction, reconstruction, and routine maintenance are not concerns of this document

nation couldn't have a well established decision making model in regard to prioretizing highways for any sort of improvement (upgrading, rehabilitaion). The manual contains guidelines for multi-criteria decision making but we couldn't get any experience done by the guideline since its publication time. Therefore, it is justifiable to build a multi-criteria decision making model to prioretize road improvement projects with a particaular emphasis to rehabilaitaion.

In short, developing a decision support tool is necessary. A number of multi-criteria decision making – MCDM (Diakoulaki et al. 2005) methods can be applied to address the challenge. But the most appropriate once can be Analytic Hierarchy Process - AHP (Abrishamchi et al. 2005; Chao et al. 2006) and Multi-Attribute Decision Making – MADM (Moges 2007). AHP is the simplest for the decision makers to understand despite its inability to accommodate feedback. On the other hand relative weighting of criteria is too subjective in MADM. Therefore, the combination of both methods is applied in this model to take advantage of each.

Statement of the problem

For a fair and effective decision output, a decision must either be multi-objective or multi-attribute or both. The theoretical research problem of this research, therefore, lies on necessity of establishing a multi-criteria decision making model for road infrastructure projects of developing countries. In Ethiopia a number of highways need to be rehabilitated or generally improved. Only about 12% of the highways are in good condition, more than 80% of the total road surface is unpaved, and person-road length ratio is the lowest in even in African standards. So, there is a huge demand for road quality improvement. But the government doesn't have capacity to do it in short period of time. For instance, in the last 10 years: 1997-2007, more than 50% of the expenditure on road development was funded by foreign aid. In such a big gap between the need to road quality improvement and the real financial capacity, Ethiopia lacks a comprehensive decision making model to prioritize road projects in order to benefit from effective investment of the little money at hand. This is the real research problem now prevailing in the nation and necessitates the establishment of a comprehensive multi-criteria decision making model to prioritize highway projects for any sort of quality improvement (e.g. heavy maintenance, rehabilitation, upgrading, or reconstruction).

Research objectives

The main objective of this research is to propose a full-fledged, comprehensive, and semi-automated model that can prioritize all highways of Ethiopia for rehabilitation. There are two specific objectives under this major aim. These are: to design a multi-criteria decision model for national road project prioritization and to implement the proposed model with real data of Ethiopia and see the results.

Methods and materials: General procedure of the research

Any decision making process begins with the recognition and definition of the decision problem. This task has already been done in the previous section. In this sub-section, the methodologies followed for data collection, data analysis and data presentation will briefly be explained. In Figure 01 Design of the research methodology Figure 01, the entire process starting from problem definition and ends up with a final outcome which is solution. The list of highways was first extracted from the entire road network and named as 'alternatives'. Then the criteria were set along with the measurable indicators within them. The modelling part continues with the AHP and MCDA (Moges 2007) methods which also include data transformation, weighting, and aggregation. The final outcome is the ranked list of the highways showing which of the highways should be given 1st, 2nd, 3rd, and so...on priority for rehabilitation.

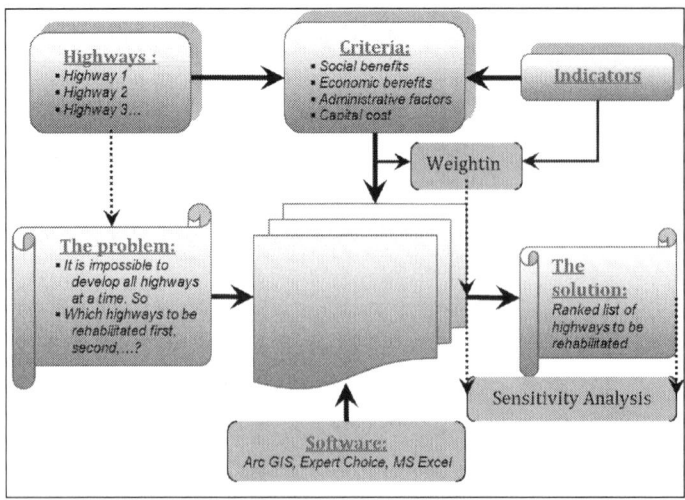

Figure 0 1: Design of the Research Methodology

ArcGIS software combined with MS Excel (Malczewski 1999) in the modelling part. Sensitivity analysis was done in order to observe the level and direction of changes in the rank of highways in response to any variation in model parameters. Sensitivity parameters in this case were: data transformation functions and weights. In Figure 01, the bold arrows indicate the major flow direction of the research task while the broken lines indicate the possible activity or step.

The multi-criteria decision modelling

This process is the core of the research and it encompasses seven major steps. These are: selecting the alternatives (i.e. highways), setting the core criteria, identifying indicators under each criterion, weighting the criteria, weighting the indicators, transforming indicator data per alternative, and aggregation and ranking. Each of these tasks is briefly explained as below.

The **first** activity was to identify the highways which compete for rehabilitation. Highways differ in their: type, size, pavement, length, and hierarchy. For instance, we can't compare second generation (Gwilliam and Kumar 2003) highways with a small local street (Riverson et al. 1991). Therefore, we followed a two-staged listing of highways and named them as: 1st order and 2nd order alternatives. Some rules were set for selecting few representative highways from the entire road network system of the country. We used two simple rules for selecting 1st order alternatives. As can be observed in Figure 02, if a highway is classified as TRUNK[4] by Ethiopian Road Authority and if it radiates from Addis to major regional city, then it is grouped as 1st order alternative.

More rules were set to select 2nd order alternatives. A road can be selected as 2nd order alternative if it fulfils at least 3 of the following parameters. These are: being a regional road, uses as motorway, has length between 50km and 800km, at least of 4th hierarchy in ERA standards, connects towns with minimum population 30,000 each, directly connects a town to international boarder, directly connects capital city of a regional state to other towns, and directly links two 1st order alternatives. Application of these filtering parameters to the entire national road network gives us 68 highways: 7 as 1st order alternatives and 61 as 2nd order alternatives. They are shown in Figure 02.

---

4 TRUNKS are those highways classified as top hierarchy according to classification by Ethiopian Roads Authority

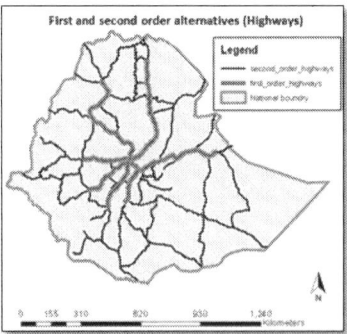

Figure 02 In total 68 highways were selected as alternatives for the decision modeling

The **second** modelling step was to set the core criteria. We referred to main issues of sustainable development in order to set our core criteria. These are: economy, social development, and environment. But since this is not a study for new projects, the issue of environment was left out and two additional criteria namely: administrative importance and capital cost were considered. Therefore, the four criteria to be used in this model are: *social benefits, economic benefits, administrative importance* and *capital cost*. By 'social benefits' we didn't mean the social issues addressed by ERA (Hine et al. 2003), rather we focus on demographic factors. More details are mentioned in further steps below. But the criteria were too crude to measure. So, we had to break them down to measurable facts.

The **third** modelling step was to identify indicators under each criterion. Measuring those 4 criteria only possible where they are further subdivided into measurable indicators as show Figure 03. A total of 22 measurable indicators were grouped under the 4 criteria. The number of the indicators under social benefits, for example (see Figure 03), economic benefits, administrative importance, and capital cost were: 6, 6, 5, and 5, respectively.

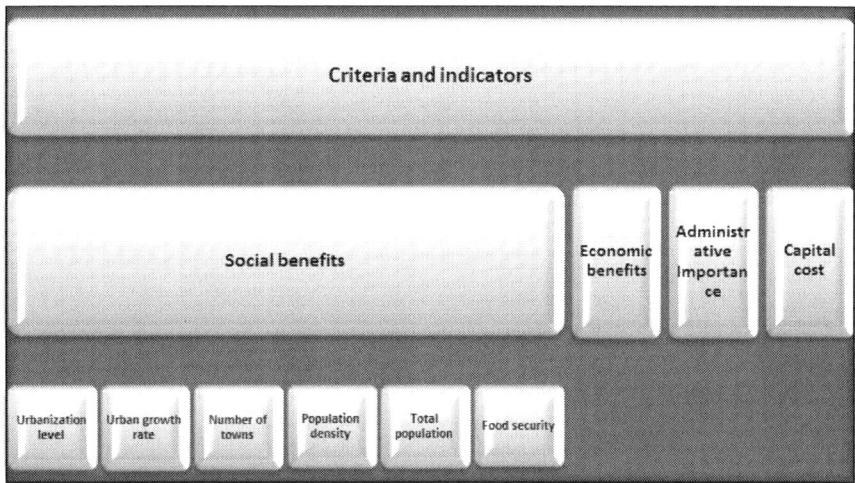

**Figure 03 Indicators under each criteria
(only indicators of social benefits are displayed)**

Indicators under social benefits were: percentage of people living in urban centres, rate of urbanization, number of towns, population density, total population, and food security. Indicators under economic benefits were access to: sites of vital national economy, mining sites, irrigable lands, export products producing areas, major crop producing zones, and livestock specialty places. Indicators under administrative importance were: connectivity to a regional capital, connectivity to sea port or boarder highway, connectivity to economically weak regions, effectiveness in road networking, and role of connecting cities. And indicators under capital cost were: road length, road surface type, topography, number of crossed rivers, and hierarchical level of the highway. Each of the 22 indicators were selected and grouped under the 4 criteria in such a way that they can be measured.

The **fourth** modelling step was weighting the 4 criteria. To do this, the procedure of Analytic Hierarchy Process (AHP) (see Figure 04) was followed. AHP is a decision making tool that was created by Dr. Thomas Saaty in 1980 (Takano, 2007). AHP allows a set of complex issues that have an impact on an overall objective to be compared with the importance of each issue relative to its impact on the solution of the problem (Álvarez et al. 2013). It uses a matrix of elements (criteria or alternatives) to make a pairwise analysis and end up with the so called "Eigen vector" which is the relative weight of the elements under consideration. The total sum of the eigen vector is 1. For our research, we used AHP in order to decide the weight of the 4 criteria. In Figure 04. The term "Ind" means indicator which is quite measurable.

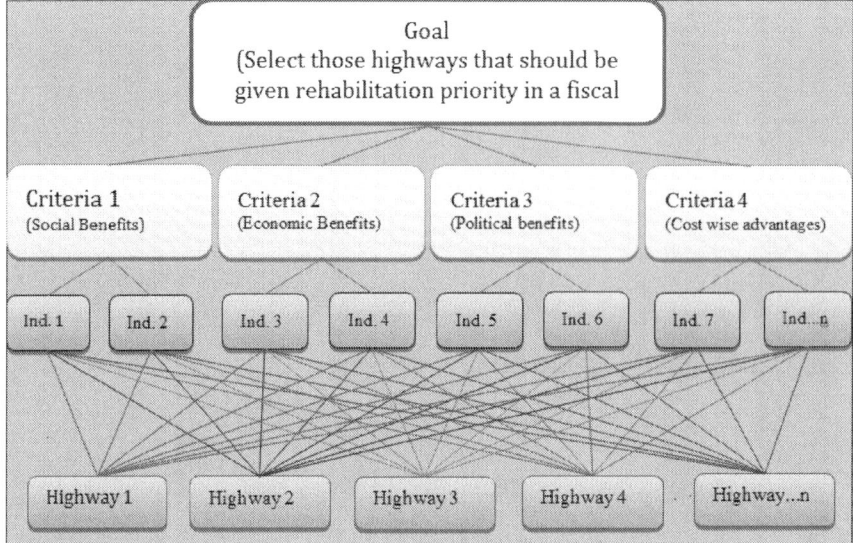

Figure 04 Application of Analytic Hierarchy Process (AHP) in highway rehabilitation

In order to set the weights of the 4 criteria, six highly educated Ethiopian experts working in transport sector were given a detailed AHP questionnaire during data collection. The analysis of the data gave us the eigen vector (which is now weight) as: 0.17, 0.38, 0.25, and 0.20 for: social benefits, economic benefits, political benefits, and cost benefits respectively. These weights are also indicated in the third column of Table 0-1.

The **fifth** modelling step was weighting the indicators. The same procedure of collecting expert judgment through AHP was applied for setting the weights of the indicators. We have to note that the sum of indicators weights within a criterion is 1 because the experts were allowed to give values for relative importance of the indicators within a criterion. The result is displayed in.

Table 01 weights of criteria and indicators from AHP

Criteria	Indicator	Criteria weight	Indicator weight
Social Benefits	Urbanization rate	0.17	0.1
	Number of Towns	0.17	0.2
	Population density	0.17	0.2
	Total population	0.17	0.1
	Food security	0.17	0.35
Economic Benefits	National organizations	0.38	0.1
	Access to Mining sites	0.38	0.1
	Irrigation potential areas	0.38	0.2
	Export producing areas	0.38	0.15
	Crop producing regions	0.38	0.3
	Livestock specialty areas	0.38	0.15
Political Benefits	Connectivity to a regional capital	0.25	0.15
	Connectivity to a seaport of boarder highway	0.25	0.40
	Connection to economically weak regions	0.25	0.15
	Level of effectiveness in Networking	0.25	0.10
	Role of connecting Cities	0.25	0.20
Cost Advantages	Road length	0.20	0.2
	Road surface type	0.20	0.2
	Topography	0.20	0.15
	Number of rivers crossed	0.20	0.2
	Hierarchical level of the highway	0.20	0.25

The **sixth** step of modelling was to transform indicator data for each alternative. This part of the research involves transformation of data of each indicator into a common format by using relevant mathematical function. We converted the real data into utility value (UV) which ranges from 0 to 1 inclusive.

One of the formulas use used to find UV is shown in Equation 2- 1, which was applied to irrigable land. Multiplying this value with 100 gives us a normalized 1 to 100 scale values, which can then be aggregated with data of other indicator transformed in a similar way.

$$UV = \frac{x_i - x_{min}}{x_{min} - x_{max}}$$

Equation 2- 1 Computation of Utility Value from real data

Where, UV is a [0-1] scale transformed value, is the area of irrigable land attached to a highway, and are minimum and maximum areas, respectively.

The last and the **seventh** modelling procedure was aggregation and ranking. This is a very important one because it wide up the overall research and gives the results we have been looking for. Mathematical expression of the final model is given by

Eq...0:1 the total score is just the summation of the sums of the four criteria. The rank of a highway (alternatives) depends on the sum the values of indicators; the larger this aggregate value the better the rank, and hence the higher priority for rehabilitation.

$$S_j = \left\{ \sum_{i=1}^{6}(0.17 * w_{isb}) + \sum_{i=1}^{6}(0.38 * w_{ieb}) + \sum_{i=1}^{5}[0.25 * w_{ipi}] + \sum_{i=1}^{5}[0.20 * w_{icc}] \right\} m_{ij}$$

Eq... 0:1 summarizing equation for the total score

Where; *Sj* is total score of highway (j), **6, 6, 5, and 5** are the numbers of indicators under each of the four criteria, **0.17, 0.38, 0.25, and 0.20** are weights of the criteria from AHP, **wisb, wieb, wipi, and wicc** are relative weights of an indicator within the respective criteria, and *mij* is transformed measure or utility value of an alternative (j) for an indicator (i).

Discussion on the results

The ultimate result of the model is the ranked list of highways. Top 20 and top 10 ranked highways are shown in Figure 01 and Figure 02, respectively. First we displayed and discussed the ranks for each criteria separately and then for all criteria.

In Figure 01(**a**), where only **social benefits** were considered as criteria, all of the 7 first order alternatives *(black lines)* are part of the top 20 highways to be rehabilitated. This is due to the reason that population and most of the urban agglomerations are located along these seven axes. The rest Majority of the rest highways *(red lines)* are connecting highlands to the desert areas in which case food security issue would be well addressed. And the model also automatically selects few highways at the central part of the country because of the high population density as compared to outer parts.

In Figure 01(b), where only **economic benefits** were considered as criteria, out of the top 20 ranked highways, 4 *(shown by black)* are the 1st order alternatives, and the rest 16 *(red lines)* are the 2nd order alternatives. The four 1st order highways shown on the map are ranked as 2nd, 3rd, 6th, and 8th. This is due to the reason that national organizations with economic importance lie along them. Moreover, most of the livestock and crop producing regions belong to these highways (e.g. Borena and Somali area).

In Figure 01(c), where only **administrative or political benefits** were considered as criteria, all the seven 1st order alternatives *(black coloured)* were automatically prioritized by the model. This tells us the vitality of those highways emanating from the capital Addis Ababa in terms of administrative advantages. Majority of out of the top 20 ranked alternatives were at the 3rd hierarchy *(they are called "Access Roads" according ERA)*. Most of these roads have intersection to the international boarder in which case political implication is justified (e.g. access to sea port and border-crossing highways).

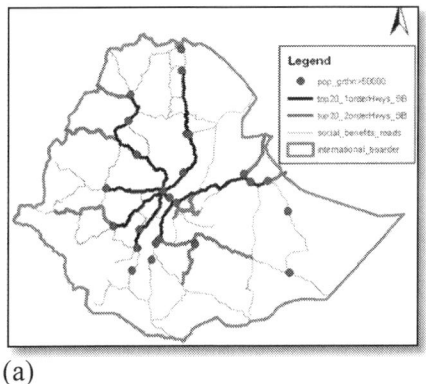
(a)

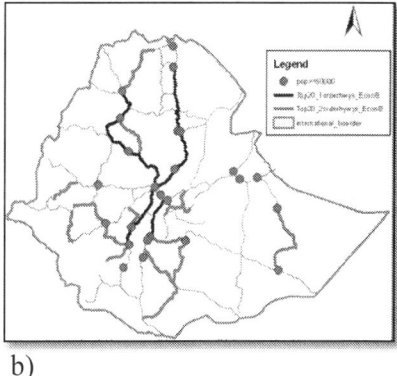
b)

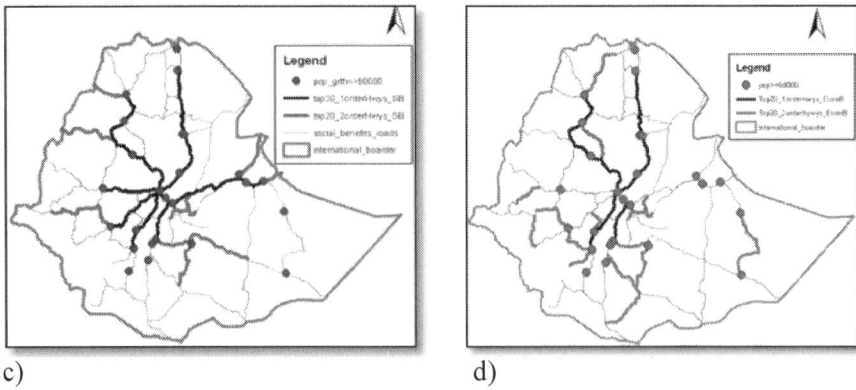

c)   d)

Figure 01 Top 20 ranked highways prioritized by using: (a) social benefits, (b) economic benefits, (c) administrative benefits, and (d) capital cost

In Figure 01(d), where only **capital cost** was considered as criteria, majority of the prioritized highways were not long which implies that road length was inversely proportional to cost. Some of the selected highways lie in the flatter parts of the country (e.g. south eastern part of the nation). This shows the influence of elevation as a factor of cost. And most of the highways with high priority for rehabilitation were of the lowest hierarchy *("Collectors" according to ERA classification)*.

The final result, which portrays only 10 highly prioritized highways, is shown in Figure 02. As can be seen from the figure, all 1st order alternatives except Addis Ababa-Hosanna-Sodo are among the top 10 ranked highways selected for rehabilitation. The result depicts that all roads are not comparable. The rest highways out of the top ten are mainly belong to regions with high crop productivity. And the purple one (in Somali Region) is prioritized mainly because of its relatively flatter landscape, oil potential and large livestock.

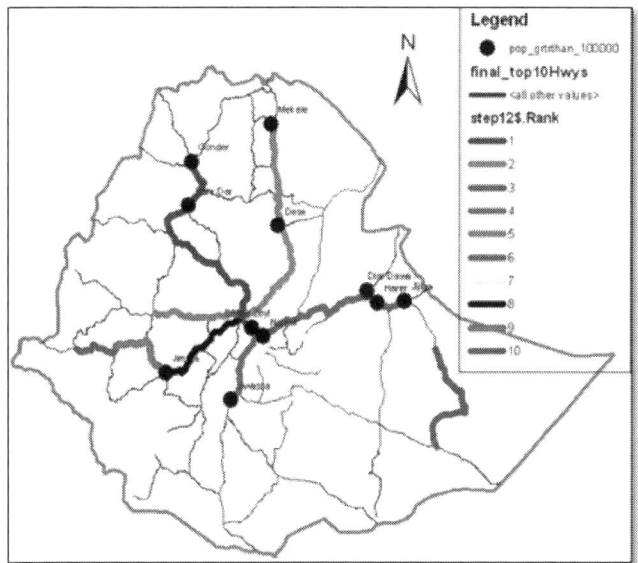

Figure 02 top ten ranked highways of multi-criteria decision making model

From the result of the model we found out that the five highways which radiate from the capital Addis Ababa got the highest priority for rehabilitation. The result, however, was highly influenced by capital cost. After removing cost as priority, regional highways got highest rank. Roads such as: Dembidolo-Gambella-Jikawo, Goba-Bitata, Metu-Gore-Gambella, Shashememe-Goba-Robe, Degahabur-Gode, and Gimbi-Asosa became: 1st, 2nd, 3rd, 4th, 5th, and 6th rank; respectively. This shows that rehabilitating these regional highways would equally be very feasible as those which radiate from Addis Ababa. The final rank of the highways was highly sensitive to changes in: data quality, transformation functions, and utility value assignment approaches. The model can be used as a prerequisite for feasibility studies.

Conclusion

Multi-criteria decision making/analysis (MCDM) is not a new concept to infrastructure planning (Modarres and Zarei 2002). But Ethiopia had never used it for highway planning until 2009. Then introduction of MCDM model to Ethiopian Road Sector Development Planning (RSDP) was necessary. This model covers the entire country of Ethiopia in which case Ethiopian

Roads Authority may use, amend and modify the model easily for all federal roads under its responsibility. Additional advantage of the model is that it may not only be used for highway rehabilitation but also for similar developments like: new construction, reconstruction, upgrading …etc. with a very little or no modification. The model is composed of 4 criteria, 22 indicators and 68 highway segments which are selected from the entire country. All the remaining highways are of low hierarchy as compared to the 68. The four criteria are social benefits, economic benefits, administrative or political importance and capital cost. Except the missing of environmental aspects, these criteria are believed to represent the concept of sustainable development.

REFERENCES

Abrishamchi, A., Ebrahimian, A., Tajrishi, M., & Mariño, M.A. (2005). Case study: application of multicriteria decision making to urban water supply. *Journal of Water Resources Planning and Management,* 131(4), 326-335.

Álvarez, M., Moreno, A., & Mataix, C. (2013). The analytic hierarchy process to support decision-making processes in infrastructure projects with social impact. *Total Quality Management and Business Excellence, 24(5-6),* 596-606.

Becker, H.J., & Demissie, B.D. (2006). Public Private Partnership in Road Projects in Ethiopia. In: IVP-Schriften, Number 9, ISSN 1613-1258.

Chao, C., Huang, Y., & Wang, M. (2006). An application of the Analytic Hierarchy Process (AHP) for acompetence analysis of technology managers from the manufacturing industry in Taiwan. *World Transactions on Engineering and Technology Education, 5,* 59.

Diakoulaki, D., Antunes, C.H., & Martins, A.G. (2005). MCDA and energy planning. *Multiple Criteria Decision Analysis: state of the art surveys* , New York: Springer; 2005. pp. 859–890.

Gwilliam, K., & Kumar, A. (2003). How Effective Are Second-Generation Road Funds? A Preliminary Appraisal. *The World Bank Research Observer,* 18(1), 113-128.

Malczewski, J. (1999). GIS and multicriteria decision analysis: New York, John Wiley & Sons.

Modarres, M., & Zarei, B. (2002). Application of network theory and AHP in urban transportation tominimize earthquake damages. *Journal of the Operational Research Society,* 53(12), 1308-1316.

Moges, F. (2007). Multi-criteria performance measurement model development for Ethiopian manufacturing enterprises. In: Addis Ababa University, in partial fulfillment of the Degree of Masters of Science in Mechanical Engineering.

Riverson, J., Gaviria, J., & Thriscutt, S. (1991). Rural Roads in Sub-Saharan Africa. *World Bank Technical Paper, 141.*

Shiferawa, A., Söderbomb, M., Sibac, E., & Alemud, G. (2012). Road Networks and Enterprise Performance in Ethiopia: Evidence from the Road Sector Development Program. In: Working Paper.

Worku, I. (2011). Road Sector Development and Economic Growth in Ethiopia. *Ethiopia Support Strategy Program II, International Food Policy Research Institute, Addis Ababa, Ethiopia.*

CHAPTER 12:

The Practice of Opting for Open Source Solutions in Higher Education Institutions of Ethiopia

Florida Alemayehu and Lemma Lessa

INTRODUCTION

As the advancement in the world of information technology and the internet escalates society has become more and more dependent on it for day to day activities, business transactions and communications. Computer technology led by the software industry has brought a huge change in the economies and life standards of countries that relay on it. Computers, mainly referring to the hardware in this sense, need software to manipulate data. Software is a set of instructions that tells a computerized hardware what to do for the achievement of a specific task. As briefly stated by (Gonzalez-Barahona, 2000), although all the stories related to software are obviously short, that of open source software is one of the longest amongst them. In fact, it could be said that in the beginning, there was only free software. Later on, proprietary software was born, and it quickly dominated the software landscape, to the point that it is today considered as the only possible model by many people. Only recently has the software industry considered free software as an option again.

Open source software is software of which the code i.e. the computer instruction, steps, procedures and techniques applied to come up with it, is freely available along with the executable product. The term open refers to the disclosure of the source code. The free tag that always comes with the name open source is not mainly about price it is meant to signify the freedom given, for those who want to and are capable of, to do any possible improvement with it.

Anyone who has these software products has the automatic license to change, customize, upgrade, redistribute or even sell it as long as the source is acknowledged (Open Source Initiative, 2010).

The origin of this software development paradigm dates back before even the emergence of the proprietary software. During the 1960's IBM released the first free modifiable source coded software with the first large scale commercial computers. Soon after, the trend of releasing both the hardware and the software by the same company has stopped and software sell became an independent industry while the hardware is manufactured by another company during the mid-1970's.

The history of open source is closely tied to the history of the hacker culture, since it is largely hackers who have sustained this movement (Bretthauer, 2001). The unbundling of the software from the hardware is one of the major reasons for the emergence of proprietary software with a commercial intent. The history of pure open source movement was begun by a guy named Richard Stallman who released the GNU and its utilities for free. In the late 1970's and 1980's the American programmer Richard Stallman working in the Artificial intelligence Laboratory at MIT, motivated by his disappointment trying to get the source code of a laser printer from Xerox in order to fix the jamming problem it had, founded the first free software initiative and the GNU project to produce an operating systems with freely available source code and the general public license (GPL) that makes sure these software stays free. Richard Stallman has written most of the tools necessary for a full operating system like the compiler and editor by the end of the 1980's but the kernel which is the core of the operating system that handles the switching of processes and the management of resources was not written by this time.

The major event happened in 1991 when a young university student Linux Torvalds wrote his own kernel based on the idea he got from an older version operating system MINIX that later became a GNU/Linux operating system under the GPL license. This action of a young researcher opened the new era for software development by becoming a reason for the emergence of a whole new community of programmers, users and implementers located all over the world that advocate, contribute and distribute this software all over the world called the open source community (Bretthauer, 2001).

The objective of this research is to reveal the current status of higher education institutes of Ethiopia towards open source adoption, to investigate the motivations behind the acceptance and rejection of OSS and to forward contextual solutions and recommendations.

LITERATURE REVIEW

Open standards, open source software and open content are issues of the same philosophy but of different interpretation and implementation. A standard in a technical context is a specification that is generally accepted and used by a specific industry. Different reasons are raised to assert the significance of open standards. According to (Nah, 2006) to employ open standards in one's company will protect it from losing data created by an application that is to be obsolete. Another benefit of using open standards is it lets the user to have multiple choices to use even to the extent of mixing products of different vendors to get the best out of all. In addition, open standards can make it easy to port one application to the other and to migrate data files very easily in case of a need for a distinct information system.

On the other hand open content is an initiative to make freely available any kind of human creation like art works, pictures, audio, video etc. for everyone interested. It follows the same philosophy and licensing as the open source software in that anyone who has this resources have the right to reuse, change or redistribute the content. The perceived usefulness of open content are improved access to information, quality of content resulted from collaborative work and peer review and improved management of content. The most famous open content practices are the Wiki and blogs. Wikipedia is an online user generated encyclopaedia that allows users to contribute, use and edit its content.

The pros and cons of both OSS and proprietary software (PS) have been addressed by different scholars. PS comes to the user in binary forms without the source code. Some (Shawn, 2005) argue this is more of letting the user use the software rather than selling it. Contrary to this OSS lets the user not only to own the source code but also the right to change, update and redistribute as long as the source is acknowledged. In terms of total cost of ownership that includes the cost of training personnel, of having associated inputs and the money needed to support and install the software in addition to the cost of having the software itself OSS has been the primary preference due to its free of charge license. Due to the hefty number of volunteer programmers from different background and expertise that are involved in bug fixing in open source software development, the time needed to find out bugs is very fast as well. This also made it more robust and bug free than a PS. Again because of the high number of people involved in writing and debugging the software, security of software from hackers and virus writers is more ensured in OSS than in a PS. Furthermore, though hiding the source code of proprietary software was meant to secure it by the producers; in practice it seems to aggravate the curiosity of virus writers and hackers to set their full attention in attacking these products specifically. Lack of ongoing reliable support and services and a personality to sue in case of software failure are among the widely mentioned drawbacks of using OSS unlike PS providers.

Open source solution is attracting too many users from the developing world mainly because of the economic advantage it offers. The absence of costs associated with supplementary licenses in having open source products is the key reason for its reduced total cost of ownership. In addition to this the ability of open source software to run on older hardware platforms and less power full computers unlike proprietary software makes it less costly and advantageous for developing countries. Many propositions have been given to put an end to the digital divide. OSS is one of the solutions proposed by many (Jamil, et al, 2008; Shiyevina, 2005) as both technological equipping and affordable way for developing countries to reach out their societal information needs. Jamil et al, (2008) have strongly emphasized the need for OSS for fast and affordable transfer of technology in terms of software, hardware and content for developing countries. According to their findings, the OSS advantage for developing countries is not only low cost but also the collaborative and networked support and reduced cost of negotiations, transactions and licensing.

According to (Sanjiva and Jivaka, 2004), open source cannot be seen as a mere product choice for developing countries, it is rather a strategic alternative for fast economic development through the reduction of cost invested for IT. According to their analysis, a developing nation can create value to its economy through IT by creating new business opportunities for private sector in the IT sector, by reducing IT cost both for government and the private sector and by improvement in the effectiveness and efficiency of governance. They put prerequisites for proper handling of OSS including the intellectual property law enforcement which avoids the common problem in most developing nations, the usage of pirated proprietary software. A country must also make sure the availability of low cost and fast internet access is for open source activities to continue in one country according to Sanjiva et al. Most of the open source communications and development is done online. So developing nations must ensure they have provided these infrastructures in universities, companies and other major sectors in advance for OSS to be disseminated and used effectively. Educational infrastructure is also among the crucial criteria to fast adoption of OSS according to Sanjiva et al. Training and teaching basic skills of using OSS must be given by higher education institutions and other educational institutions with in a developing country.

Higher education institutions with their increasing demand for information technology and tight budgets have to devise ways of cutting costs of having IT infrastructure while keeping technology as a major way to excellence. Universities and other high level educational institutes require software for their fundamental teachings like operating systems, word processing and the internet as well as for back end servers(web, proxy, file, mail, database and firewalls), computer laboratories, staff desktops, and PC's that are used for different purpose within the campus. Higher education institutions also need special purpose software such as content management system, course management system, finance system, registrar system and library system. There is a very promising possibility that these institutes can save a huge amount of money by using open source solutions (Darius, 2007; Kaan, 2006; Sanjiva and Jivaka, 2004). They can also prepare their students for a rapidly changing technological world, harness the culture of innovation through customization and further improvement of a product and create favourable condition for software independence in one country in the long run (Dean, 2000). Furthermore, they can foster the customs of free dissemination and reuse of knowledge through advocating the use of open source solutions for academic purpose.

Stephen and West (2005), on their conceptual model for OSS adoption, studied the varying strategic importance of IT for different organization to place a foundation for the question when to adopt new technology? Their framework considers the trade-off between features, risk and cost for open source adoption. They saw the degrees of IT importance for a given industry based on the type of the industry and a firm's position within the industry. ITs' importance for the firms' performance can be strategic, factory, support and turn around. By strategic it is how IT is related to the business strategy. Organizations use a range of information systems in which some are for competitive advantage and some are supportive. They insist on considering product attributes like features (what's new or valuable about a new product), risks (reliability, available support and vendor dependability) and cost (initial purchase price, ongoing support cost, upgrade fees and total cost of ownership including personnel, related equipment costs such as, power, air conditioning, security etc.) depending on the strategic importance of an IT system for the organization when adopting a new technology.

In their framework Stephen et al categorized differences of strategic importance of IT systems within a firm into strategic that give competitive advantage over rivals for which high feature to create differentiation is expected, mission critical which are firm specific systems that keep the organizations ongoing for which reliability should be the main concern to mitigate risk e.g. transaction processing, support that boost the efficiency of the organization for which efficiency and low cost are the main driver e.g. desktop applications, and laboratory systems, which are new technologies being experimented that can grow into any of the stages or can also be avoided without being used, future advantage should be the driver to keep this kind of software. So according to Stephen et al whenever adopting a new technology in order to mitigate risk one has to begin with the lowest strategic value systems. Higher education institutions in a developing nation are non-profit making especially those that are owned by the government. Therefore, differentiation is not an issue for their information system need. They can benefit from the findings of Stephen & et al in that since much of their investment on IT is concentrated on support and mission critical stages they can minimize cost by adopting open source software products that are reliable, full featured and less costly than a counter proprietary product.

METHODOLOGY

The main research strategy used for this study is descriptive survey which is useful mostly when a research objective includes the opinions, attitude and practice of few selected respondents. Both quantitative survey and a qualitative interview are implemented as a main means to collect data for to ensure the depth and width of information.

The source population for the study includes all government and private owned universities. The individuals (target groups) who were contacted for data collection at each institute are those people responsible for software procurement/purchasing, developing institutional ICT policies, overseeing implementation of ICT policies, developing/administering institutional ICT budgets, designing/approving software licensing agreements, approving software development in house, and developing ICT training. Technical staffs that are involved in software development and administration like web design, ICT teaching, training, network administration, system administration, database administration and those that give technical support are also included.

Even though there are more than twenty universities and colleges that belong to the government and too many that are privately owned in Ethiopia, due to budget and time limitations the researcher was obligated to use purposive sampling to select those universities that will give it the most useful information for the purpose of the survey. The scope of the research was to cover only universities. University Colleges, Colleges and technical schools are not included in this research. A total of six universities both from government and the private sector were selected for the study based on the information richness and easy accessibility.

In order to increase the reliability and validity of the data collected, the questionnaire for this research is adopted from a similar research held in UK which is made available online by the FLOSS World project of the University of Maastricht. Though most of the questions are adopted from the original source, some modifications and contextualization are made in order to meet local context. Because the questionnaire is well tested and implemented before there was no need for the researcher to a conduct a pilot test before using it.

Interviews were held only with the highest ICT officials of each institute except for one institution that does not have such a position in its organizational structure. In general in this research five ICT directors representing the five governmental institutions sampled for this research were interviewed. The overall theme of the interview questions was on their policies and software preferences for their organization. The main aim of the interview questions was to identify the attitude of high level ICT management towards open source software, their current stand and what they plan for the future of their organizational software need. Seven main questions were presented for interviewees. These questions cover whether or not they currently possess institutional ICT policy, weather this policy or draft policy specifies any software acquisition procedure or preference if there is one, if open source software is mentioned in any way in the software acquisition procedures, to what extent and for what purpose are they being used, what the motivation is behind open source software acquisition and what challenges are there in doing so and finally what they plan for the future.

FINDINGS: Awareness & attitude

The attitude & level of awareness towards Open Source Software(OSS) is assessed through the presence of explicit ICT policy considerations; the practice of considering it in software procurement; the level of knowledge of ICT service staff on OSS; and the belief of ICT officials on what is best for their organization.

Accordingly, there are no stated ICT policies of any kind in use by now in any of the institutions. However, it was possible to find out there are draft policies about to be approved and ratified in 3 (50 %) of the institutions. OSS is considered in the policies to be approved according to the ICT directors of the institutes. The considerations in Policy range from advocacy to simply putting it as an option. The other 3 (50 %) of the institutions do not have draft policies as well however; in practice all of the institutions (6 in number) consider OSS when procuring software.

Few ICT service staffs have extensive awareness in 83 % of the institutions (an OSS special team in 2 (33 %) of the institutions) and majorities have basic awareness or knowledge on OSS. This special OSS support team has the task of deploying, customizing and supporting open source software. Only one constituting 17 % of the total population of the institutions obligates all of its staffs to support OSS. This institution demands an OSS support experience at the time of staff recruitment. It is part of the job specification of some individuals to support OSS in 3 (50 %) of the institutions. Some individuals have the skills to support OSS but it is not part of their job specification in 2 (33 %) of the institutions. It is part of the job specification of all ICT staff to support OSS in 1 (17 %) of the institutions. On an individual scale, 82 % of the respondents think it is better for their institute to use both PS and OSS in the future while 14 % have an extreme stand that they want their institute to use OSS only.

Practice

The practice of opting for OSS software was assessed in three levels: at the Server; at the Desktop; & at other campus information system needs. Deployment of OSS software at the server was found to be very high in higher education institutions of Ethiopia. 83 % of the institutes use OSS almost for more than 90 % of their software needs on servers. 40 % of these institutes want to use only OSS on the server while 60 % of them use PS some times.

Though there are a number of server operating systems on the market, it looks like only Linux brands, Windows and Solaris are the most favourite ones in the higher education information system. The results of this question illustrate the three brands Linux, Windows and Solaris have dominated the server operating system deployment in the higher education institutions of Ethiopia. It was evident from the results that 33 % of the institutions use only Linux flavours (Red Hat, SuSE, Debian, Ubuntu, Fedora**)**, while 50 % of them use all Linux, Windows server and Solaris. As to the findings of this research Linux is used by 83 % of the higher education institutions in Ethiopia as a preferred server level platform while Windows server is being used in 67 % of the total institutions. However, the researcher has observed the special Linux brand that is famous and widely used differs from institution to institution. One of the institutions uses only Open Suse series, in the other institute Debian is the major operating system used and Fedora dominates the server platform in another institute. The remaining 17 % (one institution) use only Windows server 2003 and earlier operating systems.

Postfix and Send mail, both open source products that route and deliver electronic mail, are used by half of the institutions surveyed. The open source product Exim and Microsoft's Exchange server are not used by any of the organizations. Postfix and Sendmail have equal number (25 % both) of users in these institutions. But the majorities (50 %) of users use mail software other than the widely known mail software.

Apache web server is the most widely used web server in the global internet market. It looks like the same situation happens in the institutes surveyed for this study as well. In response for the type of web server software used in the institutions, a dramatic result was found showing all (100 %) of the institutions sampled for this research use open source web server products for their servers including the institution which indicated it only wants to deploy proprietary software. 67 % of this institutions use only Apache web server and the rest 33 % use Apache web server and Apache Tomcat server. From this result Apache web server is found being used by all the higher education institutions of Ethiopia.

The software used on database servers is also dominated by open source software just like the operating system and web server software. The finding for the type of database server used in these institutions signifies 50 % of the institutions use MySQL only as back end data base server software. 33 % of them use both OSS database servers MySQL and PostgreSQL. The rest 17 % use Microsoft SQL server only. The summary of the outcomes show 83 % of the institutions use OSS database servers while 17 % use proprietary database server software.

Moodle, an open source software product, is found to be the most familiar and only used virtual learning environment in the institutions surveyed for this research. All of the institutions (100 %), (The five governmental institutions) that indicated they use a Virtual Learning Environment (making 83 % of the total population), said they use Moodle as a VLE which is an open source software.

The operating system used on desktops unlike the server is dominated by proprietary software product Windows. All (100 %) of the institutions surveyed for this research are using Windows as a main desktop platform and 17 % of them use only Windows. 83 % of the respondents, however, said they use Linux for desktop along with Windows. The researcher observed Linux is used as desktop operating system most of the time by ICT service staffs and individuals who have a keen interest in open source software among the academic community.

Joomla (content management system), Zimbra (web mail), Koha (library system) and BIND (DNS server) are also among the commonly used OSS.

Reasons for using OSS at the server and the desktop

Better response with bug fixes, better interoperability with other products, saving on total cost of ownership, lower likelihood of getting 'locked in' by a vendor, the modifiability of the source code as needed, the possibility of migrating data across systems and the need for custom made software are the mainly deriving reasons for using open source software on servers in the institutes surveyed for this study in the order of decreasing importance.

Proprietary software is almost the de facto software on desktop in most of the institutions. Many reasons have been taken as decisive for this reality. Personnel preference, the fact that the software is already in use and so no switching costs , the need for a specialized software, performance of the software, low staff support costs and the availability of expert support are the most influential reasons for using proprietary software on desktop in decreasing order of influence.

However, the reasons for using open source software on those few desktops owned by the ICT staffs, according to the users in decreasing order of influence are ideological reasons, saving on total cost of ownership, lower likelihood of getting locked in by a vendor, the ability to modify source code, better response with bug fixes and support, better interoperability with other products and the need for custom made software.

Reasons for excluding OSS at the server and the desktop

Two methods were employed to know the very reasons for avoiding OSS. One is to find out their reasons to use PS which indirectly shows us what they lacked from the OSS and the other way is to know their explicit reasons to exclude OSS.

Generally, the most important reasons taken serious by more than 50 % of the respondents for using proprietary software on servers in these institutes are the availability of Expert support, performance of the software, personnel preference and the fact that there is no enough in house expertise on OSS, in decreasing order of importance.

Personnel preference, the fact that the software is already in use and so no switching costs, the need for a specialized software, performance of the software, low staff support costs and the availability of expert support are as well the most influential reasons for using proprietary software on desktop in decreasing order of influence.

Most of the reasons presented as a potential cause for avoiding open source software at the server are not regarded as a main barrier by the respondents. Out of ten reasons put for excluding open source software, all of the reasons are regarded as not reasons at all by more than 50 % of the respondents. In general, training needs, legal issues, interoperability and migration problems and migration cost are potential reasons for avoiding open source software.

Much of the reasons given as potential cause for avoiding open source software at the desktop have not been found sound enough for most of the respondents as well except for training needs. 53 % of the respondents believe training needs can be taken as a reason for not using open source software at the desktop in their organization.

Motivations and challenges of Using OSS among managers

The main motivations for using open source software seem to revolve around the issues of cost, security and customization among managers of the higher education institutions surveyed for this research.

Main Motivations	Main Challenges
• Cost reduction • Security from Virus • The need for customizable software • Better feature (specially on server software)	• Wide spread use of pirated software • User resistance to change • Lack of support • Lack of awareness about OSS • Lack of in house expertise and support on OSS

Table 1: Main motivations and challenges of using OSS

ANALYSIS AND DISCUSSION

The attitude and awareness among ICT officials and ICT service staffs towards open source software in higher education institutes of Ethiopia is found to be high. Even though, one of the institutions has not shown much interest towards open source software. The reason behind this can be low awareness. Awareness of open source software have not been found very high among 83 % of the institutions as well, but due to the highly skilled few individuals (OSS support team in some of the institutions) they are able to manage and support their OSS deployed on the servers. However, the awareness of open source software among the rest of community in the institutions is very low as indicated by the ICT officials, which made it very difficult to migrate to open source software at the desktop level.

From the practices of the institutions it is evident OSS has been the primary choice for server applications in most of the higher education institutions even in those that do not have policies that advocates or prefers OSS. The results of the interviews with the ICT officials on the purpose and extent of open source software have shown open source software is also used for most of the information systems in the institutions in addition to servers. However, the outcomes of this study signify the practice of using open source software on desktop is almost none in all of the institutions surveyed. All (100 %) use Windows operating system on desktop. Some open source operating systems are deployed on desktops in 83% of the institutions. These desktops are mostly used by the ICT service staffs and some interested groups only. This indicates that there is a huge difference in the adoption of open source software on servers and desktops in the higher education institutions of Ethiopia selected for this research.

The main cause for the low adoption of open source software at the desktop as described by most of the ICT officials first and for most is the wide spread use of pirated proprietary software. Due to long lived and wild use of pirated software, there are even those that do not know there is such a thing called proprietary software and that they are using an illegal copy of software. Consequently, the availability of proprietary software at no price has over shadowed the inherent advantages of using open source software for free in higher education institutions of Ethiopia as it is in many developing nations. This situation has created reluctance towards the adoption of open source software on both mangers and users. Nevertheless, the director for one of the institutions said one of the motives deriving their institution in planning to migrate to open source software on desktop level is to avoid illegal use of software with in the campus. The second reason contributing for this huge gap cited by most of the directors is the resistance to change. Users are not willing to change the status quo. They are already familiar with Windows operating system and they became dependent on it since it has been there starting from their first encounter with computers. A lot of effort is needed to change this situation which is in the future plans of many of the institutions. Lack of awareness is also mentioned by these officials as a main challenge in this regard. 53 % of ICT service staffs believe training need is the major cause of avoiding open source software on desktop as well.

The results of this study in general indicate the major obstacles for full utilization of open source software in higher education institutions of Ethiopia are mainly four, which are the use of pirated proprietary software, lack of expertise in open source software, lack of awareness and resistance to change among users. The use of pirated software had become a culture even to the extent of users not knowing that the software they are using is illegal. In addition to its illegality, this practice has overshadowed the need for a sustainable and secure solution like the use of open source software. This situation in turn caused a deeply rooted reluctance and ignorance among managers and users.

Lack of expertise in open source software is among the chief challenges for the adoption of open source software in higher education institutions of Ethiopia. ICT staffs are usually trained for proprietary software products when they are recruited. Except for one institution the rest of the institutions do not demand their staffs to have a good knowledge of supporting open source software before they are recruited. There are a group of individuals who work on open source software in most of the institutions; but they are few in number. This has caused the institutions to have less confidence on deploying open source software as much as they like.

Lack of awareness especially among desktop level users is also one of the obstacles for higher education adoption of open source software. Many of the users have not even heard of open source software let alone know what it means and what it can offer them according to ICT officials contacted for this study. Except for some computer elite individuals and ICT staffs the rest of the community has been trained computer courses using proprietary software products and that is the only software they know. Due to this reason users only want to use what they are already familiar with.

The fourth challenge and most probably the direct result of the above three difficulties is the resistance to change. As it is indicated on the reasons for using open source software in this study 91% of them said the main driver is personal preference. Users are not usually willing to change the status quo in spite of the advantages offered to them in a change.

CONCLUSION AND RECOMMENDATION

From the findings of this study the researcher believes the use of pirated software, which blocked the effective utilization and acceptability of open source software at the desktop according to this study, must be strictly banned within the campuses and the country in general and long term solutions and strategies must be planned instead. Solutions that are less costly, long lasting and dependable like the open source initiative must be implemented at every level and in all sectors especially in the academic sector as soon as possible. Effective handling of Intellectual property right can also be possible through the promotion of open source software to bear the consequences of such laws. Training staffs with the necessary knowledge to support open source software must also be the next very critical task for higher education ICT officials in Ethiopia. As the findings of this research indicate lack of expert support is found to be the main barrier for not adopting open source software in most of the institutions. One way to alleviate this problem can be by demanding the ability to support open source software at the time of new staff recruitment and by creating appropriate atmosphere for already existing staffs to have a good acquaintance and interest to open source software.

Even though the awareness of high ICT officials and ICT staffs on open source software is found to be good in most of the Universities according to the results of this research, there is a lot to be done when it comes to convincing higher managerial officers and the community in general to use open source software at their personal computers. Frequent trainings and awareness creation campaigns must be mobilized by the higher education institutions both within their own community and the society in general in order to alleviate the problem of resistance to change and dependency on one kind of vendor specific product. Open source software must be included in the curriculum of every student to prepare future human power that will serve the government and other sectors in this regard. The roll of higher education institutions in awareness creation, training and skills development in OSS is crucial. A mobilized effort involving all public sectors, private sectors and the non-profit making sectors should be launched to raise public awareness and usage of open source software in the education sector and the country in general.

ICT policies must be prepared and put in place in all of the higher education institutions of Ethiopia for a better and organized management of ICT resources and proper handling of capacity building activates like awareness creation to be undertaken. Policy makers in general and higher education ICT policy makers specifically must take into consideration the benefits of open source software in reducing cost in their decision making process and must focus on software that are most relevant in the local context. In most of the higher education institutions, ICT consumption reliability and stability of software is more preferable than functionality and high rated features. Open standards and formats should be formulated and stimulated by a governmental body or a multi sector task group and be published for all companies and sectors to follow on their ICT development and acquisition endeavours for a better interoperability, innovation, long term access to data and avoiding dependency. Every aspect of software development within higher education institutions as well as within the government, especially e- government initiatives, must seriously consider the open standards, open format and open source issues in advance before any kind of investment. Prerequisites for proper utilization of open source software like low cost and fast internet access and intellectual property law enforcement must also be incorporated within the policies.

Finally, the opportunity provided for future software independence and local innovation through open source software should be seriously considered by higher education research and development officials and the government. Higher education institutions in combination with the government can create a favourable environment for open source software research and development initiatives. Further research can be held to extend and enhance this research especially in terms of scope for better generalization. This research has revealed the current status of open source software adoption in higher education institutions of Ethiopia and the factors and barriers inhibiting its adoption. The same research topic with the same methodology can also be repeated to assess the situation in other public and private sectors in the country. In addition researches can further investigate the impact of building OSS human expertise on the adoption rate of OSS, what factors affect the process of software selection in the public sector of Ethiopia, the degree of dependency on proprietary software in the country and its future impact, open standards and open formats knowledge and attitude as well as on the feasibility of open source software policy for Ethiopia based on the findings of this research.

REFERENCES

Bretthauer, D., (2001). Open Source Software http://digitalcommons.uconn.edu/libr_pubs/7 [Accessed 29 June 2010].

Darius, H., (2007). Gaining competitive advantage in a knowledge-based economy through the utilization of open source software. *The journal of information and knowledge management* systems, 37(3), 284-294.

Dean, K., (2000). "Open Source Opens Education," *Wired News*, [Available at]: http://www.wired.com/news/culture/0,1284,34807,00.html

Gonzalez-Barahona, J. (2000). A brief history of open source software http://eu.conecta.it/paper/brief_history_open_source.html, [Accessed June 28, 2010].

Jamil A. Mohab A. and Hamid N. (2008). Open Source: The next big thing in technology transfer to developing nations. *International Association for Management of Technology*, IAMOT 2008.

Kaan, E. (2006). Economical and social benefits of free and open source software/Report. July, 2006. D09 Version 2.7, *OSSad*, TUBITAK-UE-KAE.

Nah S. H. (2006). *Free/Open Source Software Open Standards.* International open source network, United Nations Development Programme – Asia-Pacific Development Information Programme (UNDP-APDIP) 2006, India.

Open source versus proprietary software: a discussion. Available online at matthewbarr.co.uk

Proprietary vs. open source. John Carroll, *Technology News*. [Available at]: http://i.zdnet.com

Sanjiva, W and Jivaka, W. (2004). Open Source in Developing Countries. January, 2004, Art. no. Open Source Initiative "The Open Source Definition" version 1.9 by in *OpenSource.Org*, http://opensource.org/docs/definition.html>, [Accessed June 29, 2010].

Shawn S. (2005). Open Source versus Commercial Software: Why Proprietary Software is here to Stay. *Inform IT network.*

Shiyevina A. (2005). The theory of FOSS and its acceptance in developing nations. *University of Goteborg*, Sweden.

Stephen K. K., West, J. W. (2005) "A Conceptual Model for Enterprise Adoption of Open Source Software" THE STANDARDS EDGE: OPEN SEASON. Ed. Sherrie Bolin. Ann Arbor, Michigan: Sheridan Books, 51-62.

CHAPTER 13:
Assessing IT Governance Practices in Ethiopian Financial Institutions

Mengistu Bogale

INTRODUCTION- Background of the Study

Information systems (ISs) are the lifeblood of any large business. As in years past, computer systems do not merely record business transactions, but actually drive the key business processes of the enterprise. In such a scenario, senior management and business managers do have concerns about information systems (Sayana, 2002; Brown, 2006; Gruttner *et al.* 2010). This fact has been conceptualized some two decades ago since the 1980s as Rockart (1982) has indicated that the role of the computer in major organizations and the tasks of its chief executives have shifted significantly where the technical oriented information systems executive of the 1960s and 1970s is rapidly being replaced by a managerially oriented executive. This also has brought a shift from centralized data processing to distributed data processing. Richard *et al.* (2008) and Mullineux (2006) described the unique nature of Commercial banks (CBs) stating that they are Financial Institutions that are key providers of financial information to the economy. The same concept also applies to insurance and microfinance institutions as parts of financial institutions. There is evidence that well-functioning CBs accelerate economic growth, while poorly functioning CBs impede economic progress and exacerbate poverty. All the above points stress that banks are the most important source of external finance and thus play a key role in allocating capital and the corporate governance (CG) of non-financial firms. Further, they are at the core of payments systems and so systemic crises are very costly.

Mullineux (2006) also claimed that banks are special because their managers have a fiduciary duty to (more risk averse) depositors as well as (more risk prone) shareholders and thus a solution to the "principal-agent problem" aimed at maximizing shareholder value is inappropriate. Gruttner *et al.* (2010) claimed that more than a way to create competitive advantage, IT plays a fundamental role in the banking market and IT Governance provides tools to manage IT structures and processes in order to appropriately support the business strategy. Implementing new IT Governance in financial institutions may be very challenging, especially when technical literature has not many examples in developing markets. Hence, the overall governance of banks and their IT governance in particular would be of critical interest for academia as well as practitioners.

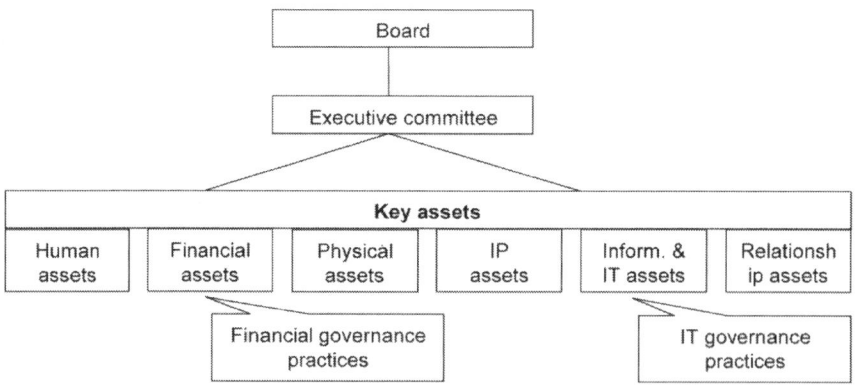

Fig. 1: The Assets Firms govern to create value (taken from Weill and Ross, 2004)

In the today's complex business environment, Weill and Ross (2004) identified six key assets namely, human, financial, physical, intellectual property, IT, and relationships that must be governed to create value.

Hence, while IT (to be used synonymously with IS in a loose sense in this paper) is one such key resources that needs to be governed for organizational value creation as shown in the diagram above.

Simonsson and Johnson (2006b) defined IT governance as the preparation for, making of and implementation of IT related decisions regarding goals, processes, people and technology on a tactical or strategic level. They developed this definition claiming that a shared view on the definition of IT governance is lacking and practitioners do not use present IT governance frameworks to support their decision-making. They also stressed that a commonly agreed upon definition of IT governance would be very useful and would serve the development and refinement of IT governance frameworks and assessment methodologies. Weill and Ross (2004) stated that information and IT are the least understood of the key assets in an enterprise. These resources being least understood will have significant negative impact on organizational survival and success. This is evidenced from the research of Parent and Reich (2009) who have reported on the importance of IT investment in the today's business environment as well as related risks due to failure of these investments. They cited as an example that U.S. companies spend as much on information technology each year as they do on offices, warehouses, and factories combined and they also claimed that IT represents about two-third of all capital spending and the average enterprise's IT investment is now greater than 4.2% of annual revenues (Parent and Reich, 2009). As a result of these large investments, the consequences of any disasters are likely to be profound and lasting.

Anecdotal evidences are available that Ethiopian Companies are spending huge amount of money on IT as indicated for the developed economies such as the USA. For instance, almost all banks are being networked through some type of banking systems so that customers will access their accounts being everywhere. Investments on ATM and other advanced banking technologies are also appearing. For instance, the June 30, 2009 audited balance sheet of Commercial Bank of Ethiopia indicated an investment of Br 120,899,950 on computers and accessories out of a total property and plant investment of Br 802,425,000 which represents some 15% of total fixed investment. In a typical business organization, investments on desktop and laptop computers, printers, telecommunication networks, database management systems and some specialized information systems (accounting, manufacturing, human resources, customer relationship, supply chain, etc.) are commonplace. As business complexity and regulatory requirements grow, the need for and investment on IT will be expected to increase. But whether these investments are really creating the value they are intended to create and whether risks related to such investments and operations are properly managed do not seem well assessed.

To strengthen this doubt, Senft and Gallegos (2009) argue that as high-speed information processing has become indispensable to organizations' activities, from a worldwide perspective, IT processes need to be controlled and from a historical standpoint, much has been published about the need to develop skills in the field of IT audit and control. This is because when information systems or technology fail, they often cause significant impacts on shareholder value (Parent and Reich, 2009). They cite the following cases as evidences for their argument. For example, early in 2001, Canada's second-largest grocery retailer, Sobeys Inc., abandoned an $89 million SAP implementation, taking an after-tax charge of $49.9 million, or $0.82 per share. Sydney Water, a public utilities company in Australia, abandoned a customer relationship management and billing system in 2002, with an estimated write off of AUD$61 million. This issue is related to information security and the privacy of the victims demanding proper IT audit and governance solutions. Based on the above facts and other similar circumstances, Parent and Reich (2009) stressed that there are ample evidences in the developed economies like the USA to suggest that just as IT can create corporate wealth it can also destroy it as 30-40% of information systems (IS) projects experience escalation, with cost overruns averaging 43-189% and also as 80-90% of all IS investments fail to meet their performance objectives as a result of senior managers' inability and/or unwillingness to exercise oversight over technology projects. This is one of the governance requirements of the board or executive management.

In the Ethiopian environment, there are anecdotal evidences indicating both negative as well as positive consequences of IT audit and governance practices. For instance, the database system of a private bank failed to retrieve transactions of some seven days' worth 10 billion Birr and customers were unable to deposit or withdraw their sums to or from their accounts. Worst of all, this has happened because the database administrator has left the organization and joined another private bank due to compensation reasons who was kind enough to come back and recover some three days data where the rest should be re-feed from paper source documents to the database through the relentless effort of employees working day and night. Personal discussion of the researcher with a board member of another private bank indicated that a decision to implement a computerized banking system has been cancelled three times after millions of Birr was paid for consultants. Another similar IT governance problem is the state owned giant commercial bank's media announcement of an employee misappropriating some 1.6 million Birr from customer deposits manipulating their ATM passwords abusing his responsibility to manage the customers' accounts.

All these facts can be taken as IT audit and governance problems according to Sayana (2004). Contrary to these negative experiences, the researcher's case study (unpublished) interviewing the Payment Card Department's manager of Dashen Bank S.C. indicated that strict control and audit through segregation of duties and proper supervision enabled it to make the ATM as secured as possible. All the above facts indicated that there are considerable differences in how IT is being governed in Ethiopian financial institutions though there are no recorded data driven studies.

Parent and Reich (2009) stressed that they see no sign that the pace and size of IT investments will abate, nor that IT-based risk will cease to be a problem. They mentioned a Computer Crime and Security Survey which found that organizations remain reluctant to report security breaches because of a concern over the effect of negative publicity. Paradoxically, these same organizations also report believing that they do not invest enough in security, including cyber-insurance. This underscores the importance for Boards to act and act now if their companies are to minimize the effects of an ever-growing array of potential IT disasters which means a call for proper IT audit and governance. Unfortunately, many Boards pay little, if any, attention to IT investments, and they do not concern themselves with minimizing potential waste or risk in this area. It can be claimed that Ethiopian Company boards and executive committee members can't be exceptions from those in the developed world.

All these problems and concerns are related to the process of IT governance in business organizations. As Weill (2004) stated, IT Governance matters because it influences the benefits received from IT investments. One of the most common and convincing reasons for the need for governance within IT is the frequent failure of IT services and projects to meet the organisation's requirements. Increasingly, there is also a level of dissatisfaction with IT's ability to innovate. Weill (2004) further claims that through a combination of practices (such as redesigned business processes and well-designed governance mechanisms) and appropriately matched IT investments, top performing enterprises generate superior returns on their IT investments (up to 40% greater return than their competitors for the same investment).

The IT Governance Institute (2007) emphasized that there is an increasing demand from boards and executive management for generally accepted guidelines for decision making and benefit realization related to IT-enabled business investments. Hence, the many stakeholders interested in IT governance that need to work together to achieve a common business goal include:

- *Board and executive*: How do we define business direction for IT, implement appropriate IT governance practices, and ensure that value is delivered and IT-related risks are mitigated?
- *Business management*: How do we define business goals for IT to ensure that value is delivered and risks are mitigated?

- *IT management*: How do we deliver IT services as required by the business and directed by the board?
- *IT audit*: How do we provide independent assurance on value delivery and risk mitigation?
- *Risk and compliance*: How do we ensure that we are in compliance with policies, regulations and laws, and new risks are identified?

One basic IT governance tool that needs critical observation is IT auditing (Woda, 2002) being a supporter of IT governance. Pathak (2005) described IS auditing (taken synonymously with IT auditing) as having acquired pre-dominance with the extensive use of information and communication technology in the business information processing area. IT auditing is, therefore, defined as the process of collecting and evaluating evidence to determine whether an information system safeguards assets, maintains data integrity, achieves organizational goals effectively and consumes resources efficiently.

IT auditing may take two forms as evidenced from the general auditing literature-internal audit and external audit. Internal audit service is obtained from company employees who are responsible for audit committees and who will report to the executive committee of the organization whereas external IT audit is independent audit service by independent public accountants and who will report to the shareholders general assembly. The contribution of both types of audits for IT governance has been indicated by Parent and Reich (2009) as shown in the diagram below:

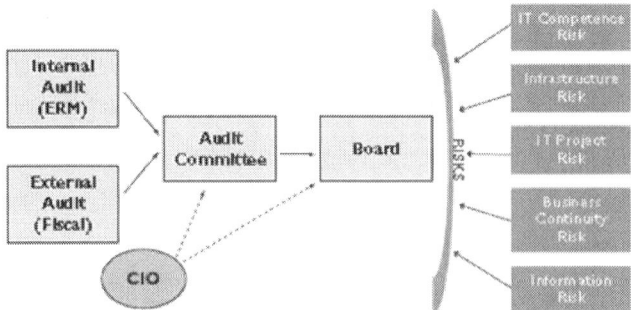

**Fig. 2: IT audit's role in IT risk governance
(taken from Parent and Reich, 2009)**

Ideally, the audit committee triangulates input about IT risk governance: the organization's Chief Information Officer (CIO), the external auditors, and internal auditors. If the CIO is relied on exclusively as a key informant, this is similar to a board telling the Chief Financial Officer (CFO) that there is no need for financial statements believing that they trust him/her when reporting that the organization is financially sound. This triangulation process is equally applicable to other components of IT governance (strategic alignment, value delivery, performance management and resource management).

However, Woda (2002) stated that the pace of change and amount of resources invested year after year in IT makes the management of IT so complex activity arguing that control in such turbulent and dynamic environment is more than a challenge (rather an adventure). Control in here means audit and assurance of IT resources, projects and operations. It is now more important than ever that the auditor become involved in supporting and helping implement corporate governance in IT and management. Africa (2009) also stated that the goal of effectively auditing and assessing the effectiveness of IT governance initiatives of a business is to allow the formulation of meaningful recommendations to management.

In developing economies like Ethiopia, there are no recorded studies about auditing of IT governance and auditing processes and their impact on organizational performance. Such studies are not available even at African level let alone Ethiopia as evidenced by Mbarika *et al* (2005) indicating that research with a focus on Sub-Saharan Africa (SSA), a major region within the world's second largest continent, is almost non-existent in mainstream information systems areas. The financial industry which is a sensitive one is not an exception. Therefore, this study has been designed to fill the gap stated in here and contribute new insight both to the academics and the practice of IT auditing and governance and its impact on organizational performance.

Statement of the Problem

The research problem to be addressed in this proposal is investigating how Ethiopian financial institutions are handling the IT Governance agenda and how IT auditing can be used in IT governance. This will be followed by assessing how IT governance performance influences performance of the institutions in Ethiopia.

The specific questions to be answered are:

1. How much cognizant are CEOs, CIOs, other C-level executives, boards of directors and Internal Auditors about their Company's IT governance (problems)?

2. How are Ethiopian financial institutions dealing with IT governance problems?

3. What is the association between IT governance performance and performance of institutions?

4. Do the Ethiopian financial institutions have IT auditing procedure to ensure IT governance process?
5. What model or framework shall be used for IT Governance process in Ethiopian context?

OBJECTIVES OF THE STUDY

The primary objective of the study is to assess the IT governance maturity level of Ethiopian financial institutions and the contribution of IT Auditing to this purpose. Based on the survey, various types of analysis will be made concerning IT governance (for instance the impact of IT governance performance on organizational performance, identifying the factors that contribute to higher level IT governance performance, etc.). Finally, the survey findings will be used for developing an appropriate model or framework for IT governance in Ethiopia context incorporating IT auditing as a means of ensuring compliance with the model or framework and other organizational policies.

Specific objectives of the study include:
1. assessing how board members and executive management are handling the different IT governance related problems in Ethiopian context- IT governance decisions, archetypes, implementation mechanisms
2. making analysis of IT governance maturity level of the Ethiopian firms
3. observing the impact of IT governance performance on the performance of financial institutions
4. assessing the role of IT auditing towards IT governance efforts
5. developing an appropriate model or framework that integrates IT auditing with IT governance for use by Ethiopian context and other countries with similar setup

Research Methodology - Target Population

The research will be conducted on a survey basis on financial institutions in Ethiopia selecting some banks, insurance companies and microfinance institutions in Addis Ababa. The list of these companies will be available from the National Bank of Ethiopian who licenses and supervises them.

Sampling Method

The sample selection will be based on simple random sampling following some cluster creation. First the target population will be categorized into two categories based on year of establishment (aged and recent ones). A simple random selection on a lot system of two institutions from each cluster will be taken as a sample for the study. A total of 12 institutions will be taken (2*3*2=12). In each institution, 10 to 15 board and executive committee members will be included in the study. This results in some 120 to 180 respondents to be in the sample. This sample size will be quite representative to conclude about IT governance process of financial institutions in Ethiopian environment.

Methods of Data Collection

Both quantitative and qualitative data will be used in the study to better substantiate the findings of the research. Such an approach has been used by Weill and Ross (2004). Others also used each option independently. For instance, Parent and Reich (2009) used the qualitative approach when investigating IT risk governance where as DeHaes and VanGrembergen (2009a, 2009b, 2010) used the quantitative approach when assessing IT strategic alignment and its relationship with organizational performance. Hence, as a logical starting point, primary data will be gathered using questionnaire prepared and included in the appendix about the IT Governance decisions, IT archetypes being used, IT governance implementation mechanisms and contribution of IT auditing for IT governance. The questionnaire used by IT governance Institute for global IT governance survey will be adapted using IT audit and governance issues stated in the literature review for the survey. To test its relevance in the Ethiopian context, the instrument will be pilot tested on small scale basis taking some two respondents from each bank in the study. Secondary data such as recorded performance data (profit, investment on long term assets, balance sheet information) will be obtained from the organizations' websites, minutes of meetings and annual reports. Such data will be used for assessing the relationship between IT governance performance and organizational performance (Weill and Ross, 2004). Following the baseline survey findings and after identifying issues that need detailed further analysis, qualitative data in the form of structured interview or focus group discussions with CIOs of the selected institutions (12 interviews or 3 group discussions having 4 members each) will be held to substantiate the findings of the survey. Based on the scale used by Parent and Reich (2009) the business organizations will be identified as green (high performers), yellow (average performers) and red (low performers in terms of their IT governance performance based on the focus group discussions).

Methods of Data Analysis

The data collected will be analysed using appropriate methods of analysis for each category of data. Quantitative data will be analysed using statistical packages such as Statistical Packages for Social Sciences (SPSS). Correlations and regressions analyses and others as appropriate will be used to observe how the variables in the study are interrelated and impacting one another. This results in IT governance maturity level for each organization, performance in each IT governance component (value delivery and risk mitigation primarily) and at last, how IT governance performance affects organizational performance. Qualitative data will be analysed by evaluating the responses of the subjects of the study using tabulations, narrations, comparisons and other types of analysis as appropriate. First, categories will be developed for each variable, then, where necessary, sub-categories will be developed. The sub-categories will be developed for the latent (unmeasured) variables. Codes will be developed for each sub-category and the interview data will be coded and assigned to the categories.

Next, the coded data for each respondent will be organized according to the appropriate sub-category and category, and these data will be used to assign an answer to each question used to measure the variables. Finally, each interview question will be given a range of possible answers and those answers for all respondents will be summarized and displayed. Based on the analysis of the survey and interview results, what model will be appropriate for customization to the Ethiopian context, what type of modifications shall be made and how IT auditing shall be incorporated in the new model will be determined based on the whole analysis of the data collected.

THE RESEARCH MODEL

To explain IT governance, the three factors identified as critical by various experts (VanGrembergen and DeHaes 2009a for example) are the following:

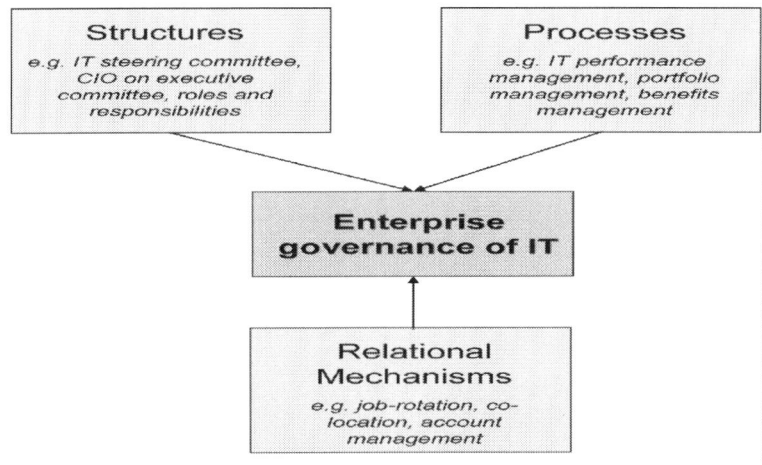

Fig. 3: IT governance structures, processes and relational mechanisms (VanGrembergen and DeHaes 2009a)

Once the factors that facilitate IT governance and the theories that explain this are identified, the relationship between IT governance performance and organizational performance can be observed as evident from the following model:

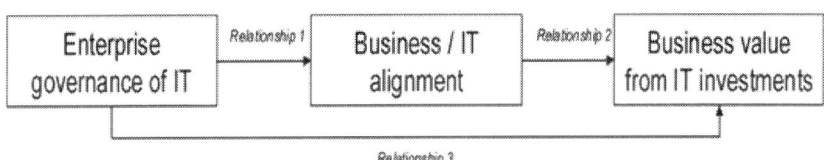

Fig. 4: IT governance Vs organizational performance (VanGrembergen and DeHaes 2010)

The model presented for Business/IT alignment and business value from IT investments (organizational performance) will be equally applicable for other components of IT governance. Even if the above model can be used to related IT governance performance and organizational performance, it doesn't indicate the role of IT auditing in the relationship. However, there are several theoretical and empirical evidences indicating this. Next, theoretical foundations that indicate the role of IT auditing I IT governance will be cited so as to develop the research model for this dissertation.

Auditing IT Governance

Hardy (2009) argued that the significance of information and IT is all around us in every aspect of business and public life and the need to derive more value from IT investments and manage an increasing array of IT-related risks have never been higher. Business executives and managers are the users and decision makers who need more value and reliability from IT and must live with the results of IT-enabled change. They now realize that they cannot abdicate their responsibilities and can no longer stand aside and leave important decisions to technicians. As sponsors and users, they are taking ownership and making organizational changes to create a more effective structure for overseeing and monitoring IT-related goals and issues. These beginnings are precursors of auditing IT governance practices in organizations. Africa (2009) stated that auditing IT Governance deals with the audit approach and procedures in reviewing IT governance processes within a business firm. It aims to show the critical areas of IT governance as well as their effects on the quality of IT service delivery to satisfy business objectives.

Based on the above model, the following theoretical model will be used for this research.

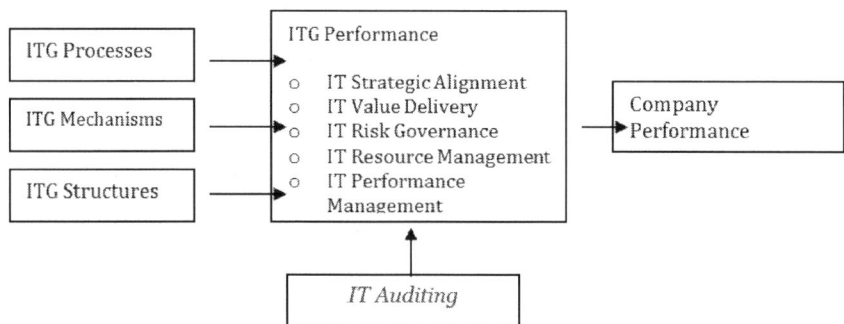

**Fig. 5: The proposed research model
(from VanGrembergen and DeHaes, 2010; Jewer, 2009)**

To strengthen the above argument, Ford and Rosam (2003) suggested a strategic change in the partnership process from the traditional auditor-business partnership. The following diagram will show this change in the role of auditors in business process management which fundamentally incorporates IT.

REFERENCES

Africa, D., (2009). Auditing IT Governance Seminar, ISACA Manila Professional Development Centre, ISACA Manila Chapter.

Brown, W.C., (2006). IT governance, architectural competency and the VASA, *JOURNAL OF Information Management & Computer Security,* 14(2), 140-154.

De Haes, S. and Van Grembergen, W., (2009a). Exploring the relationship between IT governance practices and business/IT alignment through extreme case analysis in Belgian mid-to-large size financial enterprises, *Journal of Enterprise Information Management,* 22(5), 615-637.

DeHaes, S. and VanGrembergen, W. (2009b). IT Governance: Moving From IT Governance to Enterprise Governance of IT, *ISACA Journal,* 3.

Ford, C. and Rosam, I. (2003).Process Management Auditing for ISO 9001:2000,The High Performance Organization, PIMS Digital, Essex, London.

Gruttner, V. et al. (2010). IT Governance Implementation – Case of a Brazilian Bank, Proceedings of the Sixteenth Americas Conference on Information Systems, Lima, Peru, August 12-15, 2010.

IT Governance Institute, (2007). *IT Governance Implementation Guide,* Second edition, USA.

Mullineux, A. (2006). The corporate governance of banks, *Journal of Financial Regulation and Compliance,* 14(4), 375-382.

Parent, M. and Reich, B.H. (2009). Governing Information Technology Risk, *California Management Review,* 51(3), 134-152.

Richard, E. et al. (2008). Credit risk management system of a commercial bank in Tanzania, *International Journal of Emerging Markets,* 3(3), 323-332.

Rockart, J.F. (1982). The Changing Role of the Information Systems Executive: A Critical Success Factors Perspective, *Sloan Management Review,* 24(1), fall.

Sayana, S.A. (2002). The IS Audit Process, *Information Systems Control Journal,* 1(1), 20-21.

Sayana, S.A. (2004). Auditing Governance in ERP Projects, IT Audit Basics, Vol. 2., http://www.isaca.org, [Accessed January 10, 2010].

Senft, S. and Gallegos, F. (2009). Information Technology Control and Audit, Third Edition, USA, CRC Press.

Simonsson, M. and Johnson, (2006b). P. Assessment of IT Governance- A Prioritization of COBIT - KTH, Royal Institute of Technology. Working Paper No. 151.

VanGrembergen, W. and DeHaes, S. (2008). Enterprise Governance of IT, Antwerp University, Belgium. Idea Group Publishing.

Weill, P. (2004). Don't Just Lead, Govern: How Top Performing Firms Govern IT, MIT Sloan School of Management, Centre for Information Systems Research, Working Paper No. 341.

Weill, P and Ross, J.W. (2004). IT Governance: How Top Performers Manage IT Decision Rights for Superior Performance, USA, Harvard Business School Press.

Woda, A. (2002). The Role of the Auditor in IT Governance, *Information Systems Control Journal*, Volume 2.

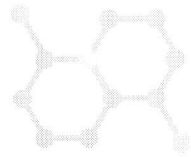

CHAPTER 14:

Hybrid Software Architecture Design Pattern Model

Asebe Jeware & Nassir Dino

1.
INTRODUCTION

The emergent of architectural patterns with some different qualities and purposes follows selection of architectural patterns in designing software architecture. These architectures have their own significant contribution for software development but most of the time one architectural pattern may not be sufficient in designing the whole aspects of system attributes, especially when it is used to implement program architecture that encompasses different projects in the enterprise. In order to solve this, there should be a systematic approach to hybridize different software architectural patterns that we can implement on designing the appropriate architecture for enterprise wide program architecture.

The rest of the paper is organized as follows. Section 2 covers the background. Section 3 presents related works to identify the gap in prior works. Section 4 discusses the proposed solution, and finally in Section 5, the work will be concluded with some concluding remarks.

2.
BACKGROUND

Hybridizing means a process of bringing two or more different things into one by mixing them together. In this paper we consider architectural patterns as a thing to be hybridized in order to create a hybrid architecture which can maximize bringing of a solution for recurrent problems in software development.

As Hassan (2009) discussed, the notion of a pattern is "geared toward solving recurring problems in design." But a pattern is more than just a battle-proven solution to a recurring problem. The proposed solution involves some kind of structure which balances these concerns, or "forces", in the manner most appropriate for the given context. So a definition which more closely reflects its use within the patterns community is:

Each pattern is a three-part rule, which expresses a relation between a certain context, a certain system of forces which occurs repeatedly in that context, and a certain software configuration which allows these forces to resolve themselves.

In this paper, pattern as an element in the world is applied to implement a relationship in a certain context, a certain system of force to resolve a problem which occurs repeatedly in that context and a certain spatial configuration by hybridizing two or more architectural patterns within single software architecture.

3.
RELATED WORK

In this Section different related literatures are reviewed in order to find the gap between what is done and what is expected to be done.

As noted by Leonardo, Vincenzo and Alessandro (1997), an Information Filtering system of HTML/Text documents is collected from the World Wide Web (WWW), based on a representation of user interests as inferred by the system through a dialogue. One of the distinguishing features of the system is the use of a hybrid approach to user modeling, in which case-based components and an artificial neural network are integrated into one coherent system. Moreover, in order to perform an accurate filtering, the system takes advantage of semantic networks and a well-structured database.

As Muhammad, Susan and Dirk (2010) stated, modern WSNs consist of a number of wireless sensor devices sometimes called *"nodes"*. A node can be a low-end or a high-end device with different storage capacities, processing capabilities, and software architecture. These nodes form networks to collect sensory data and transfer it to a base station or sink, sometimes referred to as gateway. The data can be used by a number of different applications managing and monitoring the network. In certain environments, these networks also comprise of actuators forming Wireless Sensor and Actuator Networks (WSANs). Open Framework Middleware (OFM) is a distributed, service orientated model driven system. However, the core issue facing OFM comes from the node level where nodes limit the function of what OFM is intended for, i.e., using models to define WSN at all levels of the network.

The research has highlighted the limitations associated with the traditional layered protocol stack approach and has shown that it restricts network software modularity. The rigidity of the layered protocol stack approach and the lack of information sharing between protocol layers impede optimal network performance as shared layer information is a prerequisite for network optimization in wireless heterogeneous environments.

For OFM to fulfill its claims, hybrid architecture is proposed which removes the stack based protocol layers and places emphasis on running a service oriented OFM micro middleware over the device abstraction level. However, it is fully concerned and implemented on network service and wireless connectivity enhancement.

Dongyan, Sunil, Catherine and Heung-Keung argues that in order to distribute video and audio data in real-time streaming mode, two different technologies - Content Distribution Network (CDN) and Peer-to-Peer (P2P) - have been proposed. However, both technologies have their own limitations: CDN servers are expensive to deploy and maintain, and consequently incur a cost for media providers and/or clients for server capacity reservation. On the other hand, a P2P-based architecture requires sufficient number of *seeds* supplying peers to *jumpstart* the distribution process. Compared with a CDN server, a peer usually offers much lower out-bound streaming rate and hence multiple peers must jointly stream a media data to a requesting peer.

Although the implementation of this hybrid architecture is successful on streaming media distribution, it is not developed considering software and software architectural patterns.

4.
THE PROPOSED SOLUTION

The proposed model is produced with the intention of implementing more than one architectural pattern within one system architecture in order to have highly strong system design which can implement all system requirements.

The hybrid architecture model will bring significant change for architects and developers, and minimize development time in having efficient and effective system architecture.

It is created based on the ISO-9126-1 quality attributes specification as the first and priority to proceed with this research steps.

It is listed 1 up to 8 as follows:

1. Analyze the main functional requirements and nonfunctional requirements for the system, to establish the quality requirements and quality goals.
2. Use the customized ISO 9126-1 quality model for the architecture as a framework. Some of the metrics could be further specified, according to specific components and/or connectors.
3. Present the initial candidate architectures based on project nature, quality requirements, and architectural patterns capability to solve quality requirements.
4. Construct the comparison table for the candidate architectures.
5. Prioritize the quality characteristics taking into account the system's quality requirements and quality goals. The customization of ISO 9126-1 to the problem domain can be used to organize hierarchically the characteristics.
6. Analyze the results summarized in the table, according to the given priorities obtained in step 5
7. Select the initial architecture, among the evaluated candidates, on the basis of the previous analysis.
8. If a finer analysis is required, scenarios or profile-based approaches could be used, considering only the quality characteristics relevant to the problem domain, obtained in step 5. Once the aforementioned activities are completed as indicated, we need to follow the steps listed below to hybridized best fit architectural patterns. The steps listed starting from 9 up to 12 are the result of this research to hybridize different architectural patterns.
9. Based on step 7, check the selected architecture is best fit (if the pattern is recommended for that application) architectural pattern that can implement most of (if more than 50% of the quality requirements are implemented) the requirements specified and recommended to be applicable to that specific system from pattern database.

10. Identify the weaknesses that are found in the selected architectural pattern.
11. Select additional architectural patterns from the list which have the capability to give solution for the weakness that is found in the best fit architectural pattern.
12. Hybridize them taking into consideration the strength in which they are very effective (recommended) in the system in the way they can work together.

To consider one architectural pattern as best fit pattern, the following criteria should be fulfilled. The requirements specified and the pattern attributes strength should be aligned (most of the requirements specified must be addressed by that architectural pattern). The phrase, 'most of the requirements' is used to show the number of requirement specifications in percentage, which is approximately attainable above fifty percent of the lists. The reason why 'fifty percent' is given, is because of different architectural patterns capability to implement the system requirements is up to the level fifty percent is similar, but above that threshold, we can find only one architectural pattern.

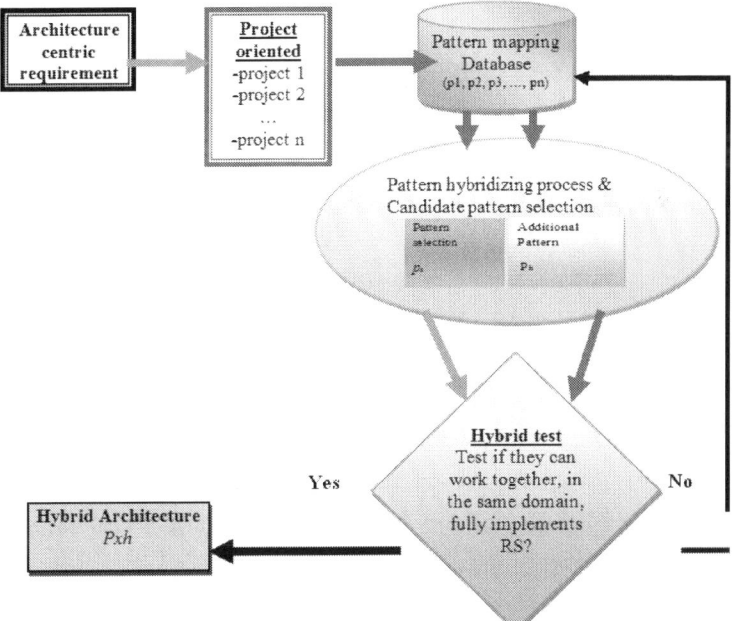

Figure 1: Hybrid pattern mod

4.3.1 Architecture centric requirement

The first step is concerned to identify architecture centric requirements from all listed requirement specifications. Architecture centric requirement is a requirement mainly concerned with the quality attributes of the system.

4.3.2 Project oriented architecture

The second box represents selection of architectural patterns that can fit into the project nature. The whole project is divided into different projects and different architectures will be implemented based on what the specific project demanded. In this step we will look into the applicability of the specified architecture inside that project. The architecture selected should be fully capable to implement all the projects' requirements or at least most of the system requirements.

4.3.3 Pattern Mapping Database

Pattern mapping database is the repository of all patterns with their quality characteristics and sub-characteristics, applicability where they can be best implemented and well accepted in the implementation, with their strength and weakness to indicate architects and other stakeholders how and when to use them and with recommended combination of architectural patterns or architectures that can be hybridized with specific architectural pattern, which can simplify the process of architecture selection in the work of architects and to use them together in system architecture design.

4.3.4 Candidate Pattern Selection and hybridizing process

At this stage the process of pattern selection will take place inside pattern hybridizing process circled by oval shape. At this stage we should identify the weaknesses of the first best fit architectural pattern that make it insufficient from implementing the whole requirements. The second architectural pattern that is going to be hybridized with the previous architecture should have the upper hand on weakness of the first architectural pattern quality attributes. That means it should have strong quality attributes that can solve the weakness of the first architectural pattern. Once we identify the additional pattern in which we can resolve the weakness of the first pattern, we hybridize them to implement it in the whole system architecture design. This process will continue till we finish solving all quality requirements found in the system and the weakness that are associated with the selected architectural patterns.

4.3.5 Hybrid test

Hybrid test is made to check or to test if the selected architectural patterns work together, if the second architectural pattern is in the same domain.

4.3.6 Hybrid architecture

The hybrid architecture is produced considering all requirements listed in the quality requirement and passes through the test case found in hybrid test stage.

The designed solution architecture considers additional architectural patterns and implements their strength in the manner of resolving the weaknesses found in the best fit architectural pattern. In order to select the second architectural pattern, as it is mentioned earlier, we should identify the weaknesses found in the first architectural pattern and then select the second architectural pattern which can resolve or give remedies for most or all of the weakness found in the first architecture.

5.
CASE STUDY

In validating the proposed solution, a case study has been conducted on one institution which is engaged on micro financing activities. In the case study, four architectural patterns have been deployed to give solution for each and every system requirements. They are OOA, SAO, N-tier, and Layered architecture which are all capable to work together.

6.
CONCLUSION

This paper focused on how to hybridize architectural patterns into the system architecture of any kind of software system without losing each pattern's respective essence. As a result, one general hybrid architectural model is produced, which can simplify the work of architects and all stakeholders; practitioners and the academia. The research uses selected architectural patterns' attributes and each characteristic of most architectural patterns and system requirements. It assists in grouping quality attributes and newly emerging quality attributes that should be considered in system architecting.

It is not recommended to conclude that architecture is all about hybridizing and to implement hybrid architecture for all types of software development. It is very efficient if we use it for big multi-project software development and for complex systems like core banking systems and huge machine controlling systems.

REFERENCES

Abdelkerim, A. and Mourad, O. (2009) *Systematic Construction of Software Architecture Supported by Enhanced First –Class Connector.* France. LINA Laboratory LINA CRNS UMR 6241, University of Nantes.

Bingfeng, X., Zhiqiu, H. and Ou, W. (2010) *Making Architectural Decisions Based on Requirements: Analysis and Combination of Risk-Based and Quality Attribute-Based Methods.* In: 2010 Symposia and Workshops on Ubiquitous, Autonomic and Trusted Computing. College of Information Science and Technology Nanjing University of Aeronautics and Astronautics.

Brad, A. (2000) *Patterns and Software: Essential Concepts and Terminology.* [Online] Available at: http://www.sci.brooklyn.cuny.edu/~sklar/teaching/s08/cis20.2/ papers/ appleton-patterns-intro.pdf/. [Accessed 22 Aug, 2012].`

Denis, H. and Maritta, H. (2009) *Deriving Software Architectures from Problem Descriptions.* Universit"at Duisburg-Essen. [Online] Available at: https://www.uni-due.de/imperia/md/content/swe/papers/2009mseim09.pdf/. [Accessed 20 Aug. 2012].

Dongyan, X., Sunil, K., Catherine R., and Heung-Keung, C. (no publication year). *A CDN-P2P Hybrid Architecture for Cost-Effective Streaming Media Distribution.* West Lafayette. Department of Computer Sciences School of Electrical and Computer Engineering Purdue University. [Online] Available at: https://www.cs.purdue.edu/homes/dxu/pubs/MMCN03-enhanced.pdf/. [Accessed 11 Jun. 2012].

Geri, S. W. (2010) *A text handout Project Architect Fundamentals*, version 5.1. Available at: geri@txt.com http://www.wyyzzk.com. [Accessed 25 Aug. 2012].

Hassan, S. A. (2009) *Investigation of the relationship between software (architecture/design) patterns and quality attributes.* Australia. The Australian National University.

Joëlle, C. (1991) *Architectural Design for Users Interfaces.* Laboratoire de Génie Informatique (IMAG). France.

Kvale, S. (1996) *Interviews an Introduction to Qualitative Research Interviewing.* Sage Publications.

Leonardo, A., Vincenzo, C. and Alessandro, M. (1997) In: A. Jameson et al. (Eds.). *Hybrid Architecture for User-Adapted Information Filtering on the World Wide Web.* Rome. Dipartimento di Informatica e Automazione, Università di Roma Tre, Italy

Mary, S. and David, G. (1996) *Software Architecture: Perspectives on an Emerging Discipline.* Pittsburgh, Pennsylvania. Carnegie Mellon University.

Microsoft Corporation. (2009) *Microsoft Application Architecture Guide.* 2nd Ed. [Online] Available at: http://msdn.microsoft.com/en-us/library/ee658117.aspx, [Accessed 15 Jun. 2012].

Muhammad, S. A., Susan, R. and Dirk, P. (2010) *A Vision for Wireless Sensor Networks: Hybrid Architecture, Model Framework and Service based Systems.* In: Digital Information Management (ICDIM) fifth international conference. [Online]. Thunder Bay. IEEE. pp. 353-358. Available at: http://ieeexplore.ieee.org/xpl/login.jsp?tp=&arnumber=5664636&url=http%3A%2F%2Fieeexplore.ieee.org%2Fxpls%2Fabs_all.jsp%3Farnumber%3D5664636/ [Accessed 27 August 2012].

Plamen, P., Ugo, B. and Robert, L. (2011) *The Need for a Multilevel Context-Aware Software Architecture Analysis and Design Method with Enterprise and System Architecture Concerns as First Class Entities.* [Online] Chicago. In: Ninth Working IEEE/IFIP Conference on Software Architecture. http://ieeexplore.ieee.org/xpl/login.jsp?tp=&arnumber=5959685&url=http%3A%2F%2Fieeexplore.ieee.org%2Fxpls%2Fabs_all.jsp%3Farnumber%3D5959685/. [Accessed 25 August 2012].

Xulin, Z. and Ying, Z. (2010) *A Business Process Driven Approach for Generating Software Architecture.* In: 10th International Conference on Quality Software. Kingston, Ontario, Canada. Department of Electrical and Computer Engineering Queen's University.

PART 3:

POLICY TOOLS TO UNPACK WHAT DEVELOPING COUNTRIES CURRENTLY FACE

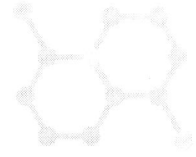

CHAPTER 15:

Exploring the extent of Information and Communication Technology (ICT) in supporting pedagogical practices in developing countries

Amare Desta

INTRODUCTION

The main focus of this paper is to identify the role ICT plays in pedagogical practices in developing nations. It also provides a framework for an on-going qualitative study into two academic institutions in Ethiopia. The review outlines what is known about ICT and related technology and their relevance and appropriateness to pedagogic practices.

Literature Review

Identifying ICT related technology

It is common to encounter the acronym ICT, especially in the education sector but what is implied by 'ICT' is by no means clear. It is therefore important to analyse the nuances of ICT and related technologies in the literature. Achacos (2003) states the term ICT often used describes the various ways in which IT related technologies are utilized/liaised in the learning and teaching process. He further argues that in the class room environment, for example, the acronym ICT is used as a descriptor for technology, and among others includes: *"technology-mediated learning, computer-aided instruction, distance education, computer-based education, technology, multimedia, communications systems, Web-based learning and computer-mediated communication ..."*

He further asserts that nature of technologies appears not to be a matter of disagreement among researchers and evaluators, as the term ICT is used by many to describe, study, and evaluate the various ways IT related-technologies integrate into education. It is also true that some authors are specific in describing specific technologies or applications; while other authors simply use the term technology to describe everything electronic.

Moreover, there is no consensus on what constitutes technology in learning or teaching. However, the common link tends to be some use of a computer based system which facilitates teaching or learning process. Though most research studies focus on computer-based technology, there are other teaching and learning technologies that are not computer-based. These may include overhead projectors, TV, VCR, DVD, sound systems, CDs, tape recorders etc. Some authors even consider the traditional piece of chalk and chalkboard a type of technology.

Newhouse (2002) claims what constitutes an educational technology is unclear. However, Rieber and Welliver (1989) defines educational technology as a process involving *"... a systematic approach to identifying instructional problems and then designing, developing, implementing, and evaluating instructional solutions."* In this paper, however, ICT will be considered as an acronym that could include any type of equipment, inside or outside of the classroom that is used to facilitate the teaching and learning process. This can include any of the non-computer based technologies mentioned above, as well as computerized technologies such as application software or Web-based learning tools.

ICT and Education

According to UNESCO's (1998) World Education Report, educational systems around the world are under increasing pressure to use the new ICTs to teach students the knowledge and skills they need in the 21st century.

The report also outlines the radical implications that ICTs have for conventional teaching and learning. It predicts the transformation of the teaching-learning process and the way teachers and learners gain access to knowledge and information. However, as argued by Kennewell (2003), quoting Rushby (1979), the need for evidence concerning ICT's positive effects on learning has been recognized since the early instances of computer-assisted learning. Nevertheless, one generation later, the evidence is still not very convincing (Kennewell, 2003).

Yet it seems that increasing numbers of educators are convinced of ICT's potential despite the inability to demonstrate clear gains from it. As indicated by Becta (1998), the utilization of digital technology within the education sector (i.e. schools, colleges and universities) is increasing as various governments tend to fund the expansion of ICT. As also stated by Teachers Training Agency (1998), various governments continue to do this mainly because of a political conviction that ICT is beneficial for all aspects of life in the 21st century and that education should be able to exploit the features of ICT in the same way as contemporary business does.

There are now indications, however, that this investment may not continue unless there is a reassurance or evidence-based justification that learners are benefiting to an extent which is commensurate with the level of provision. A significant body of interpretive research over a number of years has yielded valuable insights into the underlying processes of learning with ICT, but such studies do not directly compare the impact of different factors. It is unlikely that isolated case studies of successful teaching, or detailed interpretive analyses of learning processes, will provide convincing evidence for the political audience.

Newhouse (2002), referring to theImpaCT2 study as reported by Becta (2002), points out that in the UK this major research programme concluded that there is no consistent relationship between the average amount of ICT use reported for any subject at a given key stage (stage of UK schooling) and its apparent effectiveness in raising standards. Newhouse further asserts that while there is no direct link between using ICT and student learning, the weight of evidence now clearly shows that indirectly there can be a significant positive impact. Over the last three decades there has been an increasing amount of research conducted to investigate this impact with increasingly clearer findings of positive impacts when ICT is used appropriately. According to Newhouse, since learning is mediated through the components of the learning environment and particularly the curriculum (pedagogy and content), it is useful to start with a consideration of the potential impact of ICT use on the curriculum.

There are other sources indicating the relevance of ICTs for supporting education and training at all levels of the educational system in both developed and developing countries. Furthermore, the introduction of ICTs into the learning activities is influencing the teaching process and support landscape in a number of countries and sub-Saharan countries are no exception. This, however, could be attributed to advances in electronic messaging and computer-mediated multimedia and presentation technologies that are making a major impact on the development and provision of educational and training programmes at all levels of the educational system across the globe, including Ethiopia.

Technology and Learning

There is an ongoing debate raging among practitioners, researchers and theorists concerning the relevance, appropriateness and effectiveness of using ICT to help students learn. According to Ferguson (1997), one of the first questions that arise is whether educational technology increases student achievement. The answer to this question is undoubtedly "it depends." Ferguson (1997), quotes Ted Hasselbring, a co-director of Vanderbuilt University's Learning and Technology Centre in Nashville, USA, as saying:

> "It's kind of like asking, 'Are pencils effective?' It depends on what you are going to do with them".

Likewise, Ehrmann (1999) argues that "Technologies such as computers don't have predetermined impacts; it's their utilization that influences outcomes". This statement seems obvious but Achacos (2003) states that many institutions act as though the mere presence of technology will improve learning. They use computers to teach the same things in the same ways as before, yet they expect learning outcomes to be better.

There are also equally interesting views about the role and status of technology. Clark (1983) argues against the view that media by themselves influence the learning process. He further articulates that

> *"...media are mere vehicles that deliver instruction but do not influence student achievement any more than the truck that delivers our groceries causes changes in our nutrition"* (p. 445).

Achacos (2003) further elaborates Clark's argument and states that ".... media do not help students to learn in any circumstance" and that "the instructional method is the source of the learning ...". (p. 26). However, Kozma (1994) disputes Clark's argument and states that "... the more appropriate question was not whether ICT based technologies do influence learning, but will they influence learning".

Kozma (1994) asserts that because we have not established a relationship between computer-related technologies and learning, this should not mean that one does not exist. He believes that, since we do not fully understand the relationship between media and learning, we have yet to measure and justify it, and the failure to establish this relationship is caused in part by our theories of learning or, more specifically, behaviourism, with its basic assumption that a stimulus causes a response. Learning, in his view, is defined as "an active, constructive, cognitive and social process by which the learner strategically manages available cognitive, physical and social resources to create new knowledge by interacting with information in the environment and integrating it with information already stored in memory" (p.8). Thus, in Kozma's view, since the definition of learning has evolved to embody more of a constructive process, our measurement of this process must evolve as well.

The above contrasting views are bound to force us to think - who is right? However, as a practitioner, this author will argue that ICT is just a tool but learning is still something that is performed by the individual. This author also agree with Clark that what the educators have to do is integrate the appropriate instructional method into his/her lesson and learning may take place accordingly.

Ruthven and Hennessy (2002) analysed the pedagogical ideas underpinning teachers' accounts of the successful use of computer-based tools and resources to support the teaching and learning of mathematics. They found that mathematics teachers were familiar with "a transmission view of teaching" but they point out that the use of ICT did help to develop more interactive pedagogy.

In another article, Osborne & Hennessy (2003) assert that that it is not appropriate to assume simply that the introduction of such technologies (i.e. ICT) necessarily transforms education. Using science education as an example, they describe the critical role played by the teacher in creating the conditions for ICT-supported learning through selecting and evaluating appropriate technological resources and designing, structuring and sequencing a set of learning activities.

They outline a *"Pedagogy for using ICT effectively"* in science education in the UK, characterizing it as including:

- ensuring that use is appropriate and 'adds value' to learning activities;
- building on teachers' existing practice and on pupils' prior conceptions;
- structuring activity while offering pupils some responsibility, choice and opportunities for active participation;
- prompting pupils to think about underlying concepts and relationships;

creating time for discussion, reasoning, analysis and reflection;
- focusing research tasks and developing skills for finding and critically analysing information;
- linking ICT use to on-going teaching and learning activities;
- exploiting the potential of whole class interactive teaching and encouraging pupils to share ideas and findings;

The authors further identify and discuss a number of reasons for using ICT appropriately in science teaching and learning including; expediting and enhancing work production; increasing the currency and scope of reference and experience; supporting exploration and experimentation; fostering self-regulation and collaborative learning and, finally, improved motivation and engagement.

ICT and the educators

Light (2009) in his studies of six schools in Chile, India, and Turkey identifies four dimensions of changes that are emerging to support more project-based and ICT-rich activities in the classroom:
- changes in teachers' knowledge, beliefs, and attitudes;
- changes in how students engage with content;
- changes in relationships among students, teachers, and parents; and
- changes in the use of ICT tools to promote students' learning.

He further asserts that when is effectively integrated into a high-quality learning environment, researchers have demonstrated that it can help deepen students' content knowledge, engage them in constructing their own knowledge, and support the development of complex thinking skills (Kozma, 2005; Kulik, 2003; Webb & Cox, 2004). He further argues that ICT use alone cannot create this kind of teaching and learning environment. Educators must know how to structure lessons, select resources, guide activities and support this learning process; many traditionally-trained teachers are not prepared to take on these tasks in a new context. He refers to the work of Bransford, Brown, and Cocking (2000) and states that, to use technology effectively, the pedagogical paradigm needs to shift toward more student- learning. This shift is not trivial or easily accomplished, particularly in countries with teacher-centred educational traditions.

The above author, along with Manso (2006) and Pérez (2006), also refers to a number of factors he believes will help teachers to integrate ICT and to support students' increased use of ICT tools for learning. He identifies teacher knowledge, time, access to ICT tools and the alignment of ICT use with pedagogical goals as the main factors.

In an article entitled "why should teachers be given training to promote effective use of ICT" Dick and Reynolds (1998) point out a number of reasons why ICT staff development has become a focus of much attention in education and they identify the following points:

- constant rapid changes in technology lead to skills being quickly outdated ;
- pupils will be living in an information-based society & need appropriate skills;
- the use of ICT in schools is being held back by inadequate teacher expertise;
- the majority of serving teachers entered the profession before the advent of ICT; and
- many new skills are involved with adopting ICT into teaching - both mechanical and pedagogical.

Furthermore, Davis (1997) argues that educators must be given sufficient support to enable them to keep up with new developments while Scrimshaw (1997) highlights how the teacher's role will need to change if ICT is to be used effectively and he states that the educators will need both support in using new technologies and time to try out and reflect on new ways of learning.

Crawford (1997) points out that learning new ICT skills is generally a low priority for teachers. He further states that this is partly because of a high workload and partly because the value of ICT is not recognized in a particular institution. He further asserts that, if ICT use is to genuinely develop within the education sector, it must be designed to meet the needs of the staff who receive it.

ICT and education in developing countries

Various sources suggest that applications of ICTs are making "visible" changes in economic and social development and they acknowledge the fact that the education sector is at the core of learning in society.

Correspondingly, the role of ICT in schools is increasing. In some countries, ICT is now at the centre of education reform efforts that involve its use in coordination with changes in curriculum, teacher training and wider pedagogic activities.

According to Kozma (2002), countries from Singapore to Chile, and from the United States to Norway, have taken the position that the integration of ICT into classrooms and curricula can improve educational systems and prepare students for the 21st century learning society. SSimilarly, international organizations, such as the Organization for Economic Cooperation and Development (OECD, 2000) the European Commission (2000), and the G8 nations (2000) have identified the need to prepare students for lifelong learning in the knowledge economy and they assign a central role to ICT in accomplishing this goal.

Furthermore, according to the World Bank, for developing countries ICTs may have the potential for increasing access to, and improving the relevance and quality of, education. The World Bank's report of 1998 asserted that:

> "ICT greatly facilitate the acquisition and absorption of knowledge, offering developing countries unprecedented opportunities to enhance educational systems, improve policy formulation and execution, and widen the range of opportunities for business and the poor. One of the greatest hardships endured by the poor, and by many others, who live in the poorest countries, is their sense of isolation. The new communications technologies promise to reduce that sense of isolation and to open access to knowledge in ways unimaginable not long ago" (*The World Development Report 1998/99*).

However, the reality is that the digital divide still exists and is bound to create a gap between those who have access to and control of technology and those who do not. This means that the introduction and integration of ICTs at different levels and in various types of education will be a challenging task

How the above developments are affecting education reform in the least developed countries (LDC), especially in the Sub-Saharan African countries, is not clear. Research carried out by Light & Linden (2008) in developing countries relies on correlation designs to test whether variables are associated with each other or utilizes a qualitative or case study approach. Such an approach provides a detailed look into why and how ICTs may be used within educational settings to boost learning outcomes, but not whether their usage leads to desired outcomes over time.

Furthermore, various sources of evidence indicate that not all countries are currently able to benefit from the developments and advances that ICT can offer. According to Chinn and Fairlie (2006), most developing countries have substantially lower rates of computer and internet penetration than the rates for developed countries. Rates of technology use are especially low in African countries; for example, in Ethiopia there are 0.31 computers and 0.16 internet users per 100 people.

Unsurprisingly, lack of infrastructure and access to ICT-based resources is a common problem. Kozma (2002) cites the work of Larson (2000) referring to the U.S. State Department figures showing that approximately 275 million people had access to online resources at the end of the twentieth century, yet less than a quarter resided outside North America and Europe.

Larson (2000), citing UNESCO's report, states that "while over 26% of the U.S. population are Internet users, only 0.8% of the Latin American population are Internet users. The figure for Southeast Asia is 0.5%, for Eastern Europe, 0.4%, for Sub-Saharan Africa, 0.1% and for South Asia, 0.04%. According to Larson, not having access is the most obvious problem. Furthermore, developing countries, such as sub-Saharan Africa, are most likely to face other challenges in using ICT to improve and reform education, such as challenges related to teacher preparation, curriculum, pedagogy, and assessment.

In 2005 *info*Dev[5] created a series of "Knowledge Maps" to outline what is and is not known about ICT use in education. The above series of papers indicates a large investment in ICTs by OECD countries, and increasing use of ICTs in the education sector in developing countries. The *info*Dev's knowledge maps also produce a series of useful resources aiming to translate what is known to work within the education sector. Moreover, the United Nations Millennium Development Goals (known as MDGs) state that "It is widely acknowledged that it will be impossible for many countries to meet many of the education-related MDGs by the 2015 deadline". In order to realize these goals a fast track initiative has been created to assist LDC countries, including Sub-Saharan Africa.

ICT and Ethiopia

For example in Ethiopia, like other developing countries, ICT usage is still in its infancy, in spite of the government's efforts to promote it. According to the Ethiopian Ministry of Education (MoE) the role that ICTs can play in widening access to education to a wider section of the population cannot be underestimated. ICT is therefore praised for assisting and augmenting literacy education and also for facilitating educational delivery and training at all

---

5   *info*Dev's (Information for Development) is a program of the World Bank and other international development agencies focusing on how the use of ICTs can help to combat poverty and promote opportunity, empowerment and economic growth in LDC countries, a field of activity often referred to as 'ICT for Development', or ICT4D for short.

levels and this has been acknowledged in the Ethiopian **ICT4DPlan**[6].

The above policy document states that the Ethiopian government recognizes the key role that ICTs can play in transforming the educational system and making education accessible to a greater proportion of citizens. Furthermore, in its five-year action plan for 2006–2010 the Ethiopian Ministry of Capacity Building stated that the government is committed to addressing the nation's human resource requirements in the area of technology use through the promotion of mass ICT literacy education and training and the increase in the use of ICTs in educational institutes (schools, colleges and universities), as well as implementing initiatives aimed at connecting schools and higher educational institutions to online resources including the Internet.

The document also states that the Ethiopian government has the vision of vastly increasing the numbers of students within Higher Education Institution (HEIs) and increasing the numbers of universities and university colleges in the country. It therefore intends to simultaneously expand the intake of the newly established 12 new university colleges within the same, short, timescale and this can be achieved by utilizing ICT.

In an article entitled "ICT in Education in Ethiopia" Hare (2007) states that:

> *Leapfrogging is the word most technical people would use to describe the advancement, at least in infrastructure, that has occurred in Ethiopia in less than 10 years. Even though the country may still have one of the lowest tele-densities in the continent, there are signs that this situation will soon change. The WoredaNet[7], the e-government communication backbone, developed by the Ethiopian Telecommunication Corporation, is a promise and a major enabler for rapid ICT development in the country.*
>
> *Already the public sector and the education sector have begun to benefit from this network, and the health and agriculture sectors have been lined up for the next phase. With all this and a corresponding ICT for education policy and implementation plan, Ethiopia is set to become a model ICT user on the continent. The infrastructure seems to be falling into place and the policies and strategies are already there. The challenge now is for the government to effectively co-ordinate the implementation of the strategy.* (p. 2)

---

6 ICT4Disthe National ICT for Development Five Years Action Plan for Ethiopia [2006 – 2010] created by theEthiopianMinistry of Capacity Building

7 *WoredaNet* "Woreda" is an administrative division in Ethiopia (managed by a local government), equivalent to a district with an average population of 100,000. WoredaNet is a government network connecting more than 611 Woredas, regional and federal government offices across the country.

However, according to Ashcroft (2005), although the Ethiopian government's vision to provide the main regions and centres of population with a HEI (Higher Education Institution) is commendable, the vision needs to be rethought. She argues that "although the government effort has to be recognized, it is simply not possible to create a university or university college on a green-field site in a few months without devaluing the idea of a university and the currency of a university education". She further argues that it is better to be honest and admit that newly created institutions cannot produce Bachelor's level graduates from the very beginning, but in their early years, they can produce useful graduates at an intermediate level between TVET (Technical and Vocational Education & Training) and Bachelors programmes. This is a level that the research underpinning this study indicates is badly needed within Ethiopia.

The British Parliamentary Office of Science & Technology report of 2006 on ICT policies cited Ethiopia as an example, stating that:

> ...in Ethiopia 40% import tariffs on ICT equipment make it too costly for all but the elite. The incumbent public telecom operator has a monopoly over all telecom services. Although the number of mobile phone subscribers is growing, uptake in Ethiopia is among the lowest in Africa. About 60% of telephones and 94% of the 6,000 internet accounts are concentrated in the capital, Addis Ababa.

The above report further stated that: 'this is due to the limited telecom infrastructure, low levels of computerization outside the capital and lack of human resources'. However it stipulated that

> "the Ethiopian government's attitude to ICT may be changing, with the establishment of an Ethiopian ICT Development Authority, and changes in management of the two key telecommunications agencies". (pp 28-29)

Despite the enthusiasm reflected in the Ethiopian government policy document there are also other authors who question the lack of appropriate evidence to highlight the resources spending for technology which is not being empirically evaluated.

In an article entitled "What does research tell us about technology and higher learning?" Ehrmann (1995) states that:

> All of us wish we had good data about teaching, learning, and technology, but few institutions are doing the work to get it. That's dangerous. Technology changes quickly and unpredictably, IT budgets are large and getting larger, and money remains tight. Lacking data, faculty and administrators make big investments of time and money with their eyes closed. In today's world, it is

important to get information that helps us see what we are doing, fix problems, and document achievements.

However, in the case of Ethiopia, despite the above limitations and criticism, the role ICT can play in widening access to education to a "relatively" wider cross-section of the population has been accepted and various ICTs are now being deployed to support teaching and learning at various levels of the educational system from primary school to university level (ESDP-III[8]).

My own experience, as well as my background reading on ICT adoption in Ethiopia, led me to believe that there are various barriers, especially in the education sector, such as issues of infrastructure support, access to ICTs, training and skills development, and hierarchical social relations that determine who has access to ICTs. However, generally ICTs are considered appropriate, even though there remain concerns about the limitations identified above.

My reading further led me to believe that the implementation of ICT is occurring in a context where the cultural and institutional barriers are not well addressed. At the same time, conservative attitudes entrenched in Ethiopian academic circles question the appropriateness and relevance of ICT in a developing nation like Ethiopia, which inhibits appreciation of ICTs. There are also views that ICTs are not the most important need for Ethiopia and that people can always find a way to get along if ICT use becomes a matter of "life and death". In my opinion, the use of ICT for interactive education could lead to communication and information richness but the main stakeholders (e.g. the Ethiopian government) do not have a clear strategy on how e-learning could take place using the Ethiopian scripts and the official language.

Appropriateness of ICT in Ethiopia

However, even with the above words of caution, the Ethiopian government is pursuing the utilization of ICT in the education sector and the Ethiopian Capacity Building and the Ministry of Education have articulated various "*Action Plans*" that clearly indicate ICT as being appropriate to Ethiopia, for the following main reasons:

- ICTs are generally seen as the basic tool for survival in the next century;
- ICTs are seen to enhance efficiency in the workplace;

---

8   (ESDP-III) Education Sector Development Program III is a program Action Plan document for (2006 – 2011) produced by the Ethiopian Ministry of Education (MoE).

- there is a strong belief in ability of ICTs to increase the ease and speed of social communication;
- ICTs help university academics reach out to other professionals in other parts of the world;
- ICT usage will join Ethiopia to the global trend.

Summary

This paper provides a summary of the relevant literature from the ICT and education field, which serves as the context for the on-going study. The review centres on the role and relevance of ICT as the focus of the paper, particularly highlighting the practical and problem-centred literature in the context of developing countries education systems.

The literature suggests that ICT use is important in terms of facilitating more interactive pedagogic practices around the world. It may therefore offer promising solutions or alternatives to many pedagogy-related concerns.

A recent review of the literature by Hennessy & Angulo and colleagues (2010) entitled *"Developing use of ICT to enhance teaching and learning in East African schools* "synthesises the literature on uses of ICT in primary and secondary schools in Sub-Saharan Africa (SSA) with a particular focus on Commonwealth countries. The review throws some light on the supporting and constraining factors that influence the rapidly increasing integration of ICT in schools and teacher education in Sub-Saharan Africa (SSA).

REFERENCES

Achacoso, M. (2003). Evaluating Technology and Instruction: Literature Review and Recommendations, the University of Texas at Austin.

Adler, P.A. & Adler, P. (2008). Of rhetoric and representation: The four faces of ethnography. *Sociological Quarterly*, 49(issue number), 1-30.

Agar, M. (1986).Speaking of Ethnography. Newbury Park: *Sage Publications.*

Agar, M. (1996). The Professional Stranger: An Informal Introduction to Ethnography. San Diego: *Academic Press*.

Ashcroft, K. (2005). Analysis and Discussion of Curriculum, Resource and Organizational Issues. Higher Education Strategy Centre. Retrieved March 24, 2010, [Available at]: www.higher.edu.et

Avison, D.E. & Myers, M. (1995). Information Systems and Anthropology: An Anthropological Perspective on IT and Organizational Culture. *Information Technology & People* 8(3), 43-56.

Becta, (2002).*The Impact of Information and Communication Technologies on Pupil Learning and Attainment.* (ICT in Schools Research and Evaluation Series – No.7): DfES.

Bentley, R., Rodden, T., Sawyer, P. &Somerville, I. (1992). Ethnographically - Informed Systems Design for Air Traffic Control. ACM 1992 Conference on Computer-Supported Cooperative work: sharing perspectives, New York, 1992,pp. 123-129, Hughes et al., 1992).

Berger, P. & Luckmann, T. (1967).The Social Construction of Reality, Doubleday. NY, Garden City.

Bernard, R. (1995). Research Methods in Anthropology: Qualitative and Quantitative Approaches (second ed.). CA, *Walnet Creek*.

Boonstra, A., Boody, D. and Bell, S. (2008). Stakeholder management in IOS projects: analysis of an attempt to implement an electronic patient file, *European Journal of Information Systems*,17(2), 100-111.

Bourgois, Philippe I. (1995). *In Search of Respect: Selling Crack in El Barrio*. Cambridge,. Cambridge University Press.

Bransford, J. D., Brown, A. L., & Cocking, R. R. (Eds.) (2000). *How people learn: Brain, mind, experience, and school*. Washington, DC: National Research Council/National Academy Press.

British Parliamentary Office of Science & Technology report on ICT in developing countries (2006).Vol. No: 261. http://www.parliament.uk/documents/post/postpn261.pdf [Retrieved January 7, 2010]

Chinn, M.D. & Fairlie, R.W. (2006), "The Determinants of the Global Digital Divide: A Cross-Country Analysis of Computer and Internet Penetration," forthcoming, *Oxford Economic Papers*.

Clark, R.E. (1994). Media will never influence learning. *Educational Technology Research and Development, 42*(2), 21-29.

Clark, R.E. (1983). Reconsidering research on learning from media. *Review of Educational Research, 54*(4), 445-459.

Collins, P.H. (1991). *Black Feminist Thought*. New York, Routledge.

Corbitt, B.J. (2000). Developing intra-organizational electronic commerce strategy: an ethnographic study, *Journal of Information Technology,* 15(2), 119-130.

Corbitt, B.J. & Thanasankit, T. (2000). Using and Validating Critical Ethnographies in Information Systems Research Finding the 'Hidden' in Thai Requirements Engineering Processes. (Department of Information Systems, University of Melbourne)

Crawford, R. (1997). Managing Information Technology in Secondary Schools London and New York: Routledge.

Davis, N. (1997c). Strategies for staff and institutional development for IT in education: An integrated approach In Somekh, B. and Davis, N. (Eds.). Using Information Technology effectively in Teaching and Learning: Studies in Pre-Service and In-Service Teacher Education, London and New York: Routledge.

Denzin, N. & Lincoln, Y. (1998).Strategies of Qualitative Research. *Thousand Oaks, CA: Sage Publications.*

Dick, M. & Reynolds, A, 1998. Reflection on two case studies of staff development with small group . http://www.adastral.demon.co.uk/ma3/assignment/offl_res.htm [Retrieved February 15, 2010]

Edwards, P. & Belanger, J. (2008).Generalizing from workplace ethnographies. *Journal of Contemporary Ethnography,* 37 (3), 291-313.

Ehrmann, S.C. (1995). What does research tell us about technology and higher learning? *Change, 27(2),* 20-27. http://www.learner.org/edtech/rscheval/rightquestion.html [Retrieved January, 2010]

Ernst, D. (2002). Global production networks and the changing geography of innovation systems: Implications for developing countries. *Economics of Innovation and New Technology,* 11 (6), 497-523.

Ethiopia, Ministry of Capacity Building (2005), *National ICT for Development 5 years Action Plan* for the year 2006 – 2010.

Ethiopia, Ministry of Education. (2006). *Education Sector Development: Programme Action plan* for the year 2006 – 2011.

Ethiopia, Ministry of Education. (2005). Education Sector Development Programme III (ESDP-III): Program Action Plan.

European Commission (2000). Communication from the commission to the council and the European Parliament: The e-learning action plan – Designing tomorrow's education. Brussels: Commission of the European Communities.

Faithorn, L. (1992). Three Ways of Ethnographic Knowing. *Revision* 15(1): 23-27.

Feller, G. (2005). 'ICT Policy of Ethiopia: changing positively', *Information for Development (I4D)*, 13(6), 28-29.

Ferguson, B., (1997), Education Technology: *An extended literature review* http://www.sdavjr.davis.k12.ut.us/~brian/ferguson/pdf_files/Exlitrev.pdf [Retrieved March 25, 2010]

G8 Countries (2000). Okinawa Charter on the Global Information Society. http://www.dotforce.org/reports/it1.html. [Retrieved April 03, 2010]

Fetterman, D. (1989). Ethnography Step by Step (2nd ed.). *Thousand Oaks*: Sage.

Fine, G.A. (1993). Ten lies of Ethnography – Moral Dimensions of Field Research, *Journal of Contemporary Ethnography,*

Fine, G.A. & Holyfield, L. (1996). Secrecy, trust and dangerous leisure: Generating group cohesion in voluntary organizations. *Social Psychology Quarterly*, 59, 22-38.

Fontanna, A. & Frey, J. (2000). The interview: From structured questions to negotiated text. In Denzin, N. & Lincoln, Y. (Eds.).*Handbook of Qualitative Research*. Thousand Oaks, CA: Sage Publications Inc.

Geetrz, C. (1973). The interpretations of Cultures, New York, Basic Books.

Geetrz, C. (1988). Works and Lives: The Anthropologist as Author. Cambridge: Polity Press.

Golden-Biddle, K., & Locke, K. (1993).Appealing Work: An Investigation of How Ethnographic Texts Convince. *Organization Science, 4*(4), 595-616.

Grills, S. (1998a). Doing Ethnographic Research Fieldwork Settings, Thousands Oaks, CA: Sage Publications.

Hare, H. (2007).ICT in Education in Ethiopia. http://www.infodev.org/en/Document.402.pdf [Retrieved April, 21, 2010]

Harvey, L.J. & Myers.M.D. (1995). Scholarship and Practice: The Contribution of Ethnographic Research Methods to Bridging the Gap. *Information Technology & People* 8(3): 13-27.

Hennessy, S., Deaney, R., & Ruthven, K. (2003). Pedagogic strategies for using ICT to support subject teaching and learning: An analysis across 15 case studies. *Research Report No. 03/1*. Cambridge: University of Cambridge.

Hughes, J.A., Randall, D. & Shapiro, D. (1992). Faltering from Ethnography to Design. ACM Conference on Computer-Supported Cooperative work: sharing perspectives, New York, pp. 115-123.

Kennewell, S., (2003). Developing research models for ICT-based pedagogy, University of Wales Swansea, Department of Education – http://portal.acm.org/citation.cfm?id=857119 [Retrieved November 3, 2009]

Kozma, R. (1983). Reconsidering research on learning from media. *Review of Educational Research,* 53(4), 445-459.

Kozma, R. (1991). Learning with media. *Review of Educational Research,* 61(2), 179-211.

Kozma, R. (1994). Will media influence learning? Reframing the debate. *Technology Research and Development,* 42(2), 7-19.

Kozma, R. (2002). Qualitative Case Studies of Innovative Pedagogical Practices Using ICT. Journal of Computer Assisted Learning 18, 387-394.

Kozma, R. (2005). National policies that connect ICT-based education reform to economic and social development. *Human Technology,* 1(2), 117–156.

Kulik, J. (2003). *Effects of using instructional technology in elementary and secondary schools: What controlled evaluation studies say* (Final Report) P10446.001). Arlington, VA: SRI International.

Kuriyan, R., Ray, I. & Toyama, K. (2008). Information and Communication Technologies for Development: The Bottom of the Pyramid Model in Practice, *InfromatioOn Society,* 24(2), 93-104.

Larson, A. (2000). Remarks delivered at the Sovereignty in the Digital Age Series. Washington, D.C.: Woodrow Wilson Center.

Lee, A. & Baskerville, R. (2003).Generalizing ,Generalizability in Information Systems Research. *Information System Research,* 14(3), 221-143.

Levina, N. (2005). Collaborating on Multiparty Information Systems Development Projects: A Collective Reflection-in-Action View, *Information Systems Research,*16(2), 109-130.

Lewis, D. (2008). Using life histories in social policy research: The case of third sector/public sector boundary crossing. *Journal of Social Policy,* 37(4), 559-578.

Lewis, I.M. (1985). Social Anthropology in Perspective. Cambridge: Cambridge University Press.

Light, D., &Manso, M. (2006, April).*Educational technology integration in developing countries: Lessons from Seven Latin America SchoolNets.* Paper presented at the Annual Meeting of the American Educational Research Association. Seattle.

Light, D.,(2009). The Role of ICT in Enhancing Education in Developing Countries: Findings from an Evaluation of the Intel Teach Essentials Course in India, Turkey, and Chile Education Development Centre.

Linden, L.L. (2008). *Complement or substitute? The effect of technology on student achievement in India.* Infodev Working Paper No. 17. http://www.infodev.org/en/Publication.505.html [Retrieved November 5, 2009]

Lofland, J. &Lofland, L. (1995). Analyzing Social Setting (3 ed.). Belmont: *Wadsworth Publishing Company.*

Malinowski, A. (1932). *Argonauts of the Western Pacific: An Account of Native Enterprise and Adventure in the Archipelagos of Melanesian New Guinea.* London, Routledge and Kegan Paul.

McBride, N. (2008). Using performance ethnography to explore the human aspects of software quality, Information *Technology and People,* 21(1).

Myers, M.D. (1997). Critical Ethnography in Information Systems. In J. DeGross (Ed.), *Information Systems and Qualitative Research* (pp. 207-224). London: Chapman and Hall.

Myers, M.D. (1999). Investigating Information Systems with Ethnographic Research. *Communications of the Association for Information Systems, 2*(23).

Myers, M.D. & Young, L.W. (1997). Hidden Agendas, Power, and Managerial Assumptions in Information Systems Development: An Ethnographic Study, *Information Technology and people,* (10)3, 224-240.

Mytelka, L.K. & Smith, K. (2002). Policy learning and innovation theory: An interactive and co-evolving process. *Research Policy,* 31, 1467-1479.

Nandhakumar, J. & Avison, D. (1999). The fiction of methodological development: a field study of information systems development, *Information Technology and People,* 12(2).

Newhouse, P (2002); Literature Review The IMPACT of ICT on LEARNING and TEACHING A literature review by Dr. C. Paul Newhouse for the Western Australian Department of Education.

Ngwenyama, O.K., Introna, D.L., Myers, M.D. and Degross J.I. (1999). New Information Technologies in Organizational Processes: Field Studies and Theoretical Reflections on the Future of Work, Norwell, MA: *Kluwer Academic Publishers*

Organisation for Economic Cooperation and Development (1999).Knowledge management in the learning society. Paris: OECD/CERI.OECD. (2001). *"Information and Communication Technologies and Rural,* www.sourceoecd.org. [Retrieved October 11, 2009]

Orlikowski, W. (1991).Integrated Information Environment or Matrix of Control? The Contradictory Implications of Information Technology. *Accounting, Management, and Information Technology, 1*(1), 9-42.

Osborne, J., & Hennessy, S. (2003). Literature review in science education and the role of ICT: Promise, problems and future directions. Report No. 6. http://www.futurelab.org.uk/resources/documents/lit_reviews/Secondary_Science_Review.pdf [Retrieved March 15, 2010]

Pérez, P., Light, D., Vilela, A., & Manso, M. (2003, 2006). Learning from the pioneers: A study on the best practices of the network TELAR. *Interactive Educational Multimedia, 6,* 17–39.

Player-Koro. (2010). How do different modalities of pedagogical practices within teacher education shape student teachers? An empirical study of secondary mathematics teacher education. http://bada.hb.se/bitstream/2320/5900/1/Paper_CPK.pdf [Retrieved May 1, 2010]

Prasad, P. (1997). Systems of Meaning: Ethnography as a Methodology for the Study of Information Technologies. In J. DeGross (Ed.), Information Systems and Qualitative Research. London: *Chapman and Hall.*

Rajpramukh, K.E. (2005). The Shaman in Asia Pacific Region: A Cross Cultural Study. *Man in India*, 85(1 & 2):25-40 (2005). http://www.krepublishers.com/ [Retrieved July 25, 2010]

Rieber, L. P. (1994). *Computers, graphics, and learning*. Dubuque, IA: Wm. C. Brown Communications.

Rieber, L. P., & Welliver, P. W. (1989).Infusing educational technology into mainstream educational computing. *International Journal of Instructional Media, 16*(1), 21-32.

Rushby, N. (1979): *Introduction to Educational Computing.* London, Croom Helm.

Ruthven, K. & Hennessy, S. (2002), 'A practitioner model of the use of computer-based tools and resources to support mathematics teaching and learning', *Educational Studies in Mathematics,* 49 (1), 47–88.

Sahay, S. & Walsham, G. (1997). Using GIS in developing countries: Social and Management Issues. Vienna, *UNIDO Publications*.

Sanday, P.R. (1979). The Ethnographic Paradigms(s), *Administrative Science Quarterly,* (24)4, 527-538.

Schultze, U. (2000). A Confessional Account of an Ethnography about Knowledge Work, *MIS Quarterly, 23*(1), 1-39.

Schultze, U. & Leidner, D.E. (2002).Studying knowledge management in information systems research: Discourses and theoretical assumptions. *MIS Quarterly*, 26 (3), 213-242.

Scrimshaw, P. (1997).Computers and the teacher's role In Somekh, B. and Davis, N. (Eds.). Using Information Technology effectively in Teaching and Learning: Studies in Pre-Service and In-Service Teacher Education London and New York: Routledge.

Sidky (2003).Critique of postmodernism anthropology - in defence of disciplinary origins and traditions, Lewiston, The Edwin Mellen. http://www.media-anthropology.net/coman_maoverview.pdf [Retrieved March 15, 2010]

Stanley, L.D. (2003). Beyond Access: Psychosocial Barriers to Computer Literacy, *The Infromatio0n Society*, 1, 407-416.

Suchman, L. (1987). Plans and Situated actions: The problem of Human-Machine Communication, Cambridge: Cambridge University Press.

Teacher Training Agency (1998).*The Use of ICT in Subject Teaching: Expected Outcomes for Teachers.* London, Teacher Training Agency and the Departments of Education, UK.

UNECSO (1998a). World Education Report: Teachers and teaching in a changing world. http://www.unesco.org/education/educprog/wer/wer.htm [Retrieved January 25, 2010]

Walsham, G. and Waema, T. (1994). Information Systems Strategy and implementation: A case study of a building society, *ACM Transactions on Information Systems,* 12(2), 150-173.

Webb, M., & Cox, M. (2004). A review of pedagogy related to information and communications technology. *Technology, Pedagogy and Education,* 13(3), 235–286.

World Bank (1998). Latin America and the Caribbean: Education and technology at the crossroads. Washington, DC: World Bank.

World Bank (1998), *The World Development Report 1998/99. Quoted in Blurton, C., New Directions of ICT-Use in Education.*

Wynn, E. (1991). Taking Practice Seriously, in Greenbaum and M. Kyng (Eds). Design at Work, New Jersey: Lawrence Erbaum.

Yin, R. K. (2002).*Case Study Research, Design and Methods,* 3rd ed. Newbury Park, Sage Publications.

Zuboff, S. (1988).In the age of the Smart Machine. New York: Basic Books.

CHAPTER 16:
Localization in Iranian Context: A Multi-Case Study

Mohammad-Ali Shafia, Mehdi Mohammadi, Ali-Reza Babakhan, Mohammad Ameri, Faramarz Lotfali

INTRODUCTION

The current situation of the national economy along with the imposed international sanctions against Iran has created a specific condition for the country. The growing need for the acquisition of advanced technologies in aerospace, railroad, pharmaceuticals, and so on, illustrates the requirement of relying on local and native capacities. There have been a number of technology transfer projects in the Iranian context in recent decades. However, it seems that the main goals of these projects including absorption, localization, and transfer of technology as well as innovations in production of new samples, have not been achieved to the full extent. Accordingly, there exists a dependence on technology exporting countries and imitation of their products in Iranian advanced industries as an avenue of the combat. That being the case, one is supposed to study factors contributing to localization of technology and creation of innovation to be able to adopt executive approaches. Hence, the current study aims to investigate some research projects to present a model of the factors affecting localization of technology at the firm level.

LITERATURE REVIEW

Investigation on strategies and stages of localization has been widely welcomed by East Asian scholars. Successful technological catch up experiences of Japan, South Korea, China and Taiwan on localization of importing technologies indicate that, several factors are influencing the process of localization at micro and macro levels. In his comparison of the applied approaches for localization of technology in Japanese and Chinese industries, Liu (2005) used the terms "closed acquisition model" and "open acquisition model" to refer to the predominant models of localization of technology in Japan and China, respectively (Cao et al, 2006). Applying the closed acquisition model, Japanese companies began to localize their technologies. In so doing, they increased their awareness of the know-how of these technologies through importing mature technologies, applying reverse engineering activities, and conducting concurrent research. Then, they created a free flow of knowledge and learning within the company by integration of research and development activities with all the processes of design engineering, purchase, production and marketing activities. They also formed interrelated loops among these processes - for sharing knowledge, increasing the rate of incremental innovations in processes and products- to create knowledge capturing opportunities and to accumulate knowledge in the companies. Analyzing the localization patterns of Chinese companies indicates that their applied model is different from that of Japanese. The Chinese initiated the localization affair by focusing on reverse engineering at the time that their country had low levels of technological capacity. In 1980s, some of the developed countries such as America, Japan and their European counterparts witnessed a significant growth in their importing trends of the production lines of different technologies. By the passage of time, Chinese policy makers realized an increase in their occurred technological gap with technology- importing countries.

Chinese scholars have extracted two main reasons for widening the technological gap. The first is the presence of a relatively pre-wide technological gap: until the early 1980s, the most important consumers of the imported technologies in China were big state companies that were not caring seriously to create innovation and localization of technologies. That is why the research institutes had not the chance of growing significantly in that period. These governmental research institutes had a poor relationship with industrial centres. The localization of technology rise expectation has led the increased peoples' awareness of the research institutes to create innovations, but because of their lack of consistency and integration with industrial and research centres could not meet these expectations very fast.

Second, unlike Japanese industries, Chinese technology importing companies spent very low costs for research and development (R&D). Additionally, their R&D centres had focused on repair, maintenance and control of qualities. In the same vein, most of the managers only focused on the hardware aspects of technology and ignored software, procedural and human aspects of technology. The low levels of skills and basic knowledge of Chinese companies led them enter small portions of the value chains. Thus, compared to Japanese companies, Chinese companies had limited activity levels in some areas such as construction of agencies for selling foreign products, assembly, maintenance and repair of products. Following an increase in the learning of technology and level of knowledge in universities and research centres, a joint cooperation was established among research centres, universities and industries. In addition to their use of absorptions and adaptations for localization of technology,

Chinese companies used the great mall of China, so that they could assemble foreign companies' products and satisfy their customers' needs by referring to the foreign owners of industries. Conveying the needs of the customers led to innovations in designing and manufacturing of products. Thus, innovation learning and exposition to innovative environments was possible for Chinese companies, thereby, they decreased their technological gap with other countries over the time. The importance of great mall of China in localization of technology has also been taken into consideration (Mu & Lee, 2005). After their analysis of three successful localization experiences in Chinese telecommunication industries, Mu and Lee (2005) referred to the large consumer market of China, joint venture for technological transfer and market segmentation as the main reasons for the Chinese gradual success in recent decades. Simply put, technological capability is the necessary condition for the acquisition and localization of technology.

In 1984, some of Chinese non-private companies contracted a joint venture for technological transfer with Bell Company in order to raise their technological capabilities. The purpose of this agreement was the acquisition of know-how and technical knowledge for manufacturing telecommunication products. To create a win-win condition in this agreement, the Chinese side was supposed to supply a large consumer market for the products of Bell Company and in contrast, the manufacturing and engineering facilities to be transferred to China. In the next step, considering the needs of their customers and consumers, Chinese producers tried to make some slight improvements in technological learning of the experts of Bell Company. Accordingly, they enjoyed from the knowledge spill overs of the production facilities and also the specialist forces. Besides, some Chinese companies prepared a cultural milieu to support localization of technology in their own companies.

They also focused more professionally on particular parts of the market and their needs and then, they segmented the needs of the market. This work contributed to deepening the technical learning and increasing the knowledge base required for the localization of technology and absorption of knowledge. The laws and supports of the government played a supportive role in the next step. In fact, the government supported its local producers by defining high tariffs for importing telecommunication products. After seven years, in 1991, Chinese companies succeeded in producing local telecommunication products. This evolutionary trend was continued forcefully in 1995 when Huawei private company was founded.

Figure 1 shows the followed process:

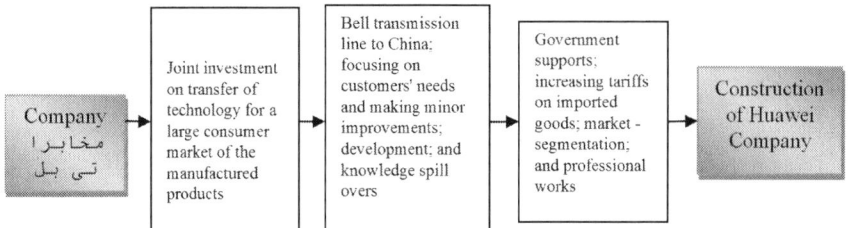

Figure 1. **The Localization Trend of Chinese Telecommunication Industries**

In another study, Choung et al. (2009) attempted to analyze the acquisition model as well as the localization of technology in South Korean semiconductor industries. South Korea began its industrial development move by establishing the assembly lines for foreign products and making use of low cost labor forces to absorb the latent technical knowledge of the manufacturing processes over time. In the next step, adopting vertical integration strategy, Korean corporate suppliers of semiconductors conquered some countries such as Japan and took over the ownership of the suppliers with the capability of manufacturing sublime technologies in these countries. Then, they moved towards mass production to take advantage of standard production and increased integration in production and specification processes of the products. Product assembly with low levels of innovation seemed to be a threat to the Korean industries; therefore, they followed architectural innovations and used advanced parts to promote performance and innovation levels of their products. They also benefited from integrating different technological components and introduced new products to raise global competitiveness.

Table 1 put together the results of the conducted studies for identifying the factors affecting the localization at micro and macro levels in the focused countries:

Table1:
Factors Affecting the Localization of Technology
(Adapted from the Previous Studies)

Factors affecting the localization of technology	Source	The Focused country
Reverse engineering; concurrent research and development with reverse engineering; allowingthe accumulationof knowledge; developing aknowledgemanagementsystem; technological learning; basic knowledge	Cao et al. (2006)	Japan
Product assembly; repair and maintenance of technologies; technological cooperation with universities and research centers; satisfying the customer needs; technological learning; human capabilities; adaptation of technology; absorption of technology	Cao et al. (2006)	China
Technologicalcooperation; culturally appropriatemilieu to supportlocalization; technologicallearning; basic knowledge	Mu & Lee (2005); He & Mu (2012)	China
Transfer of the manufacturing technology and assembly line; evolutionary development; technological cooperation; imitation of foreign technologies; innovations in imitated foreign products	Choung et al. (2009)	South Korea

In the subsequent sections, the methodology and conceptual model of this study is described. Then, the validities of the applied conceptual and final models are determined via the analysis of the obtained data.

METHODOLOGY

Based on its data collection approach and functional purpose, the pre-set study is a descriptive one of mixed method design that leans on some case studies. In this study, the common qualitative and quantitative methods were implemented to collect the data. First, factors affecting the localization of technology were extracted from the published literature. Then, observations, semi-structured and in-depth interviews, analyses of the written documents as well as a questionnaire were used to complete the list of the extracted factors and formulate a model proportionate to the Iranian context. Participants of the present study were managers, experienced engineers and experts who were present at Jihad Collegiate of Iran University of Science and Technology and TajrobehNour Turbine Company. Snowball sampling was used for the qualitative part of this study. The content validity of the questionnaire was confirmed through consultations with experts. Moreover, after piloting the questionnaire to a representative group of subjects and applying SPSS software for the data, an overall Cronbach's alpha reliability of .961 which is reliability high is obtained. Then, the developed questionnaire was distributed among all the 120 subjects of the selected sample. To model the factors affecting the localization of technologies, the data collected from 91 subjects who had answered the questions with great precision, were analysed by LISREL software and structural equation modelling approach (SEM). SEM is a statistical approach for testing the hypotheses regarding the relationships among the observed and latent variables. In SEM, the theoretical models of certain populations can be tested by correlational, experimental and non-experimental approaches (Bazargan et al, 2004).

THE CONCEPTUAL MODEL

To design the conceptual model, the recent published scientific documents were studied and analysed. From the existing documents, two most successful experiments in terms of technology abortion criteria including the localization of the electrostatic filters and turbine making were investigated by in-depth and semi-structured interviews with the present experts at Collegiate Jihad of Iran University of Science and Technology and TajrobehNour Turbine Company. In addition to the interviews, the existing documents and written records of the localization processes were analysed. Then, observations of the production workshops, machinery, assembly lines, research laboratories and R&D branches were recorded. Taken together the referred factors in the literature and the criteria considered by the present experts of the study, the following 25 factors are extracted as those affecting localization of technology at the company & firm level:

Table 2: *Factors Affecting the Localization of Technology in Iran*

Number of Factor	Factor	Descriptions
1	Learning by doing (Jensen et al, 2007)	This is the type of learning resulted from attempts for understanding the technical know-how. It contains some issues such as organizational inspection, reverse engineering activities, etc.
2	Learning by using (Jensen et al., 2007)	It is considered as one of the learning styles in companies. Its purpose is to identify different scientific and technical aspects of a technology through its application in the field of action.
3	Learning by interacting (Jensen et al., 2007)	The nature of this type of learning is different from the two previous types of learning. The main focus of this type of learning is interaction with others including suppliers, partners of strategic alliances, customers and so on.
4	Organizational Science and Technology System (Fuglsang and Sundbo, 2005)	In this type of learning, the company tries to develop a knowledge management system, provide support for the innovations and other systems that induce formation of organized scientific and technological cycles within the company.
5	Production under standards and certification of the foreign cooperators	This measure that is one of the outputs of conducted interviews refers to a form of cooperation between the companies in which a young technical certifier company attempts to investigate the technological productions of a newly established company and give it production rights.
6	Under the license agreement (Arora, 1996)	It means transfer of technology by under the license methods.
7	Reverse engineering (Radosevic, 1999)	It means acquisition of technology by reverse engineering.
8	Joint investment (Mowery et al, 1998)	It is also one of the most popular forms of technological transfer. The abbreviation used for this concept is JV.

9	Creating a place for accumulation of knowledge	This factor was considered by the interviewed experts. In their views, a central institution functioning as a documenter of experiences and knowledge of the successful local projects should be established to give a possibility for expert awareness as the source of localization in company, thereby allowing and increasing the absorption capacity of technology in organizations.
10	Flat organizational structure	According to the interviewed experts, this factor can improve the relationships among people and solve real world problems far from resolving the complex issues of the organization.
11	Managers' wills and demands	This factor is also one of the outputs of the conducted interviews. Experts believed that one of the main factors in an organization is the wills, demands and plans of the managers for localization of technology and creation of local breakthrough in technological advancements.
12	The knowledge milieu and culture of the company (Ekvall, 1996)	The culture supporting innovation and localization of technology is an important factor in the movements of company towards innovation and technology.
13	Trust between the receiver and owner of the technology	This factor creates an atmosphere of trust among the owners and receivers of technology, prevents the abuse of the technology by receivers; and observes intellectual ownership and so on, so that the technical knowledge can be transferred more appropriately.
14	Concurrent studies for identifying the problems and removing them by a foreign party	This factor is the result of the native Iranian experiences with which the suppliant of technology increases its exploitation of technological co-operation by conducting multilateral and rapid research about different aspects of technology; and also tries to achieve a large extent of technical knowledge by asking questions from the owner of the technology.
15	The internal mechanisms of the company to establish relationships among people	This factor was also the result of the expert opinions based on which the internal systems of the companies increase the flow of knowledge and absorption capacity to create a defined mechanism for enhancing communications among people.

16	Basic knowledge (Kim, 1997)	In many studies, this factor is considered as a key measure of absorption capacity. It can be concluded that the more the basic knowledge of the receivers, the more the possibility of the catch up and localization of technology.
17	The training capability for rapid application of the learnt knowledge	According to the researchers, people who have the capability of teaching can more quickly achieve the ways for applying the theoretical knowledge that they are faced with. These people play an important role in catch up and localization of technology.
18	Analysis of one's needs and answering them for discovering the extent to which the needs are met by new technologies (Cao et al, 2006)	One of the factors affecting localization of technology is the ability to adapt the technology to one's needs and make minor changes in the imitated technology. This factor suggests that if the companies have greater analyses of their requirements, they can follow the way towards catch up and localization of technology.
19	Considering local needs and adding some characteristics to the technology to meet the local needs (Cao et al, 2006)	In addition to needs analysis, the ability to identify local conditions and apply useful technological strategies for satisfying local needs is an important factor in localization of technology.
20	Initial evaluation of the overall technological performance and identification of the performance and its effects (Porter el al., 1980)	Localization and catch up of technology can be effective when the company has achieved a complete understanding of the social, economic and environmental effects of technology and is able to properly assess the desired technology.
21	The ability to apply technology	This factor which has been extracted from the conducted interviews refers to the lowest level of capabilities in localization of technology and suggests that we cannot follow the next steps towards localization, unless we can learn how to work with a technology.
22	The ability to maintain and repair the technology	According to the present experts in this study, repair and maintenance activities are one of the effective factors in formation of technological learning in the projects for catch up and localization of technology. Maintaining the old and foreign technologies, Iranian experts became familiar to the technical specifications of these technologies and achieved the capability of reverse engineering.

23	The ability to re-assemble the technology	After doing the repairs, a high level of skills is required for creating technical drawings. Expertsbelieve thatthe ability tore-assemble technologies together with theincreasinglevel oftechnologicallearning enables the company to create technical drawingsand follow assembly-procedures.
24	The ability to imitate the technology	The appropriate level of learning about technological specifications and the extraction of technical drawings lead to this step. In this step, the company is enabled to create an imitated sample of the technology.
25	The ability to create innovations in technology	After achieving the ability to create similar technological samples, widespread spectrum of different gradual innovations; innovations in components; and so on is needed. By making improvements in technology, this spectrum would increase the technological knowledge and the firms' abilities to create fundamental innovations.

After extraction of those factors affecting localization of technology based on conducted studies, the conceptual model of their relationship with localization of technology is developed as Figure 2:

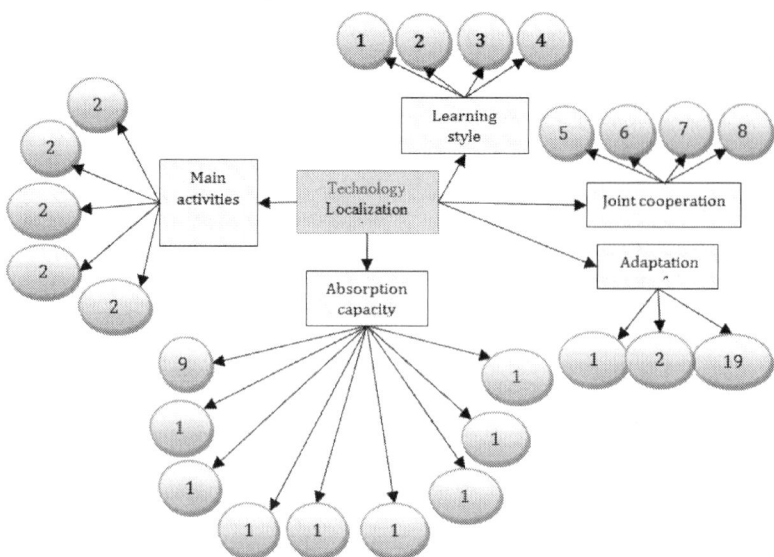

Figure 2: The Conceptual Model of Factors Affecting the Localization of Technology in Iran

DATA ANALYSIS

The SEM method and LISREL software were utilized to analyze the data collected from 91 questionnaires and evaluating different dimensions of the purposed conceptual model.

Based on the yielded results, the companies should be cautious to work on five areas to gain the ability for localizing their technologies as follows:

Learning styles: It is one of the most important elements of technological change in developing countries (Viotti, 2001). In one hand, it can be claimed that the goal of all the localization and innovation activities in companies, is an increase in their learning abilities (Tidd and Bessant, 2011). This goal can be achieved by implementation, utilization, and establishment of a scientific and technological system (Jensen et al., 2007; Fuglsang and Sundbo, 2005). The results show that all of these four learning styles have played an important role in formation of the localization processes investigated in the analysed case studies.

Technological cooperation: Any technological interaction for acquisition, adaptation, absorption and localization of technology with national & international companies as well as manipulation of their products in the form of reverse engineering is considered as a technological cooperation. Experiences of Iranian local companies show that although there are some imposed sanctions that have made the owners of international companies not willing to have technological cooperation with Iranian companies, under technology transfer license agreement is still considered as one of the common methods for transfer of technical knowledge. Experiences of the Collegiate Jihad of Iran University of Science and Technology show that, presence in the tenders of other countries requires the certification of the local built-in technologies. Although the technical approval of the foreign countries is not considered to be in the category of the common methods for transfer of technology, the findings of the Collegiate Jihad suggest that the presence of experts and specialists in the teams interacting with the warranting company can create the opportunities for absorbing the technical knowledge and solving the technological problems. It should be noted that despite the Iranian imposed sanctions, international companies are ready for conducting joint investment projects with Iranian competent companies. According to the interviewed experts, foreign companies have observed the high technological power of the Iranian companies and have decided to allot their administrative, technical, manufacturing, and commercial resources with Iran, so that they can be informed of the Iranian companies' mechanisms for localization of technology in the win-win interactions.

Technological adaptation: As the experiences of some countries like China, shows, the ability to increase the native and local requirements for foreign technologies and arrange them based on their corresponding customer requirements, underlies the localization of technology. The findings of this study not only confirm the results of the previous studies on the importance of evaluating economic and social aspects of technology, but also emphasize the importance of analysing the native requirements.

Technological absorption capacity: This factor is one of the key elements for technology localization. Technological absorption capacity can be investigated in terms of different aspects including organizational structure, organizational culture, basic knowledge, human capabilities, the level of trust among the owners and the receivers of technology, R&D capabilities, and so on. The results of this study suggested that the examined samples of the present study have managed to localize technology by promoting the level of technological absorption capacity and following the evolutionary path of technology over years. Analysis of the opinions of the present experts indicated that despite certification of the positive aspects of establishing institutions for collecting and preserving the knowledge and experiences of localization of technology, the experts do not use such institutions in their activities. This is because establishing such institutions requires the availability of necessary infrastructures such as intellectual property rights, mutual trust between providers and users of these experiences, and so on.

Main activities: This area manages those sets of technical activities that are related to localization of technology, & enable the companies and human forces to promote their capabilities from the level of working with technology to the level of innovating something new. In their initial steps towards the recognition and acquisition of knowledge of their work, the labor forces would learn how to operate and exploit the available facilities. As the depth of their knowledge about the assigned works is gradually increased, they realize new facts about the facilities and the relationships among various components of these facilities. In addition to the technology-related devices and equipment, they grow awareness of the instructions, qualifications and competencies, as well. During this process, growth and development of self, the power of solving the problems and dealing with the inadequacies in the working processes, or let's say attempts for repairing and reverting the work system to the initially defined situations are resulted. Thus, one learns to do the repairs and resolve some of the well-known software and hardware problems and becomes ready to move towards a new stage of cognitive development which is known as rectification. Based on the existing necessities, one may take advantage of one's physical abilities to repair some of the components of work or the available facilities, solve some of the hardware inadequacies, and sometimes think about using software to rectify or change the relationships governing the working system. One also might consider some criteria to move towards the enrichment of the qualities. During this process, the ability to repair is considered as a superior capability for that individual. Gradually, one becomes able to understand and imagine a more logical relationship among the components, & thinks about how and why of the deep relationships, attends the high labor standards, and finally gains more credits in his best performances. Because of the nature of practical instructions that make the one more responsible, in this stage, the person tries to fix and change the labor conditions to improve the abilities. In fact, one gets to a stage beyond his previous envisions. Quantitative values are gradually replaced by intellectual and qualitative ones and then by the flexibility, innovation and evolution as the other sets of values. In this process of growth and development, one gains the ability to distinguish and understand the relationships among various related labor components, which is a step beyond technology. An increase in one's knowledge of work, leads to an increase in one's abilities, skills and awareness of his undertaken responsibilities. The emergence of a competency to imitate and copy components from some works to the other works is the prominent feature at this stage. Then the recognition, the ability to identify the criteria and patterns are developed. This process begins a new step towards a jump in the effort values, and revives one's creativity, innovation and desire for change. In this way, an individual gets in line with the process of technological development and takes advantage of the potential capacities.

CONCLUSION

In this study, some the conducted research projects regarding the successful experiences of technology localization in some countries like Japan, China and South Korea were compared critically and the factors affecting the move in these countries were identified. Then, because of the importance of localization in the Iranian context and the significance of the analysis of the successful Iranian localization experiences, a case study was conducted to investigate the localization of filtration in Collegiate Jihad of Iran University of Science and Technology and Tajrobeh Noor Turbine Company. Different methods such as semi-structured and in-depth interviews, observations, analysis of the existing documents, questionnaires, and structural equation model were used to conduct the present study. The results indicated that affecting factors on localization of technology can be divided into five parts:

1) learning styles including learning by doing, learning by using, learning by interaction, and organization of science and technology 2) technological cooperation including under the license production and the approval of the foreign party, under the license agreement for transferring technology, reverse engineering and joint investment 3) technological adaptation including analysis of one's needs and conditions and finding some ways for meeting them by new technologies; considering local needs and conditions and adding some characteristics to the local technologies for satisfying these needs ; and initial evaluation of the overall technological performances and identification of their outputs and effects 4) technological absorption capacity including flat organizational structure, the managers' wills and demands, the knowledge milieu and culture of the company, mutual trust between the owners and receivers of the technology, concurrent projects for identifying and removing the problems by the foreign party, internal mechanisms of the company for establishing interpersonal relationships, basic knowledge and training capability to apply the learnt knowledge and 5) main activities including the ability to use the technology, the ability to repair and maintain the technology, the ability to re-assemble the technology, the ability to imitate the technology and the ability to create innovations in the technology.

REFERENCES

Arora, A. (1996). Contracting for tacit knowledge: the provision of technical services in technology licensing contracts. *Journal of Development Economics,* 50(2), 233-256.

Bazargan, A., Sarmad, Z., & Hejazi, E. (1998).Research methods in behavioural sciences. *Tehran: Agah Publishing Co. (in Persian).*

Cao, Y., Sakai, H., Liu, X. L., Nagahira, A., & Iguchi, Y. (2006, July). Technology Catch-Up in China Compared with Japan: A New Development Model. In Technology Management for the Global Future, 2006.PICMET 2006 (Vol. 3, pp. 1030-1039).IEEE.

Choung, J. Y., Hwang, H. R., & Hameed, T. (2009, December).Patterns of technology catch-up in Korean private sector. In Industrial Engineering and Engineering Management, 2009.IEEM 2009. IEEE International Conference on (pp. 93-99). IEEE.

Ekvall, G. (1996). Organizational climate for creativity and innovation. *European journal of work and organizational psychology,* 5(1), 105-123.

Fuglsang, L., &Sundbo, J. (2005). The organizational innovation system: three modes. *Journal of Change Management,* 5(3), 329-344.

Jensen, M. B., Johnson, B., Lorenz, E., & Lundvall, B. Å. (2007).Forms of knowledge and modes of innovation. *Research policy,* 36(5), 680-693.

Kim, L. (1997). Imitation to innovation: The dynamics of Korea's technological learning. (city), Harvard Business Press.

Mowery, D. C., Oxley, J. E., & Silverman, B. S. (1998). Technological overlap and interfirm cooperation: implications for the resource-based view of the firm. *Research policy,* 27(5), 507-523.

Mu, Q., & Lee, K. (2005). Knowledge diffusion, market segmentation and technological catch-up: The case of the telecommunication industry in China. *Research Policy,* 34(6), 759-783.

Porter, A. L., Rossini, F. A., Carpenter, S. R., Roper, A. T., Larson, R. W., & Tiller, J. S. (1980). Guidebook for technology assessment and impact analysis. New York, NY, Elsevier North Holland, Inc.

Radosevic, S. (1999).International technology transfer and catch-up in economic development. Cheltenham, UK, Edward Elgar Publishing.

Tidd, J., & Bessant, J. (2011). Managing innovation: integrating technological, market and organizational change, Wiley.

Tung, R. L. (1994). Human resource issues and technology transfer. *International Journal of Human Resource Management,* 5(4), 807-825.

Viotti, Eduardo B. (2001). National Learning Systems: A new approach on technical change in late industrializing economies and evidences from the cases of Brazil and South Korea. Science, Technology and Innovation Discussion Paper No. 12, Centre for International Development, Harvard University, Cambridge, MA, USA.

Zahra, S. A., & George, G. (2002). Absorptive capacity: A review, re-conceptualization, and extension. *Academy of management review,* 27(2), 185-203.

CHAPTER 17:

Knowledge Management Practices in Development and Humanitarian Aid Organizations in Ethiopia

Harmella Ayalew

INTRODUCTION

Effective delivery in the fields of development aid relies on freely flowing knowledge and its communication as well as reach of dissemination. Through Knowledge Networks and communities of practice development organizations can promote partnership, teamwork, with the free exchange of knowledge and best practices between the organization, operational partners and international agencies. Development assistance organizations, practitioners and stakeholders are considered as key actors in the production and communication of knowledge. Knowledge exchange among practitioners in development assistance is now at the forefront of global development policy formulation. This is also true in Ethiopia where a multitude of development agencies interact, produce knowledge and share among them to meet their development assistance objectives in support of the country's growth and transformation plan.

To better understand the ongoing state of knowledge management (KM) in development aid organizations, this study adopted the perspective of Kruger and Synman (2003) to assess and describe the process in which knowledge management is defined, managed, controlled and implemented in knowledge-intensive Development Aid Organizations in Ethiopia.

A defining key role development aid organizations have is building the capacity of beneficiary entities to process and use knowledge to engage effectively in substantive development debates and decision making processes (Keeble, 2002). This mandate requires development organizations to have quality learning and knowledge management systems. In addition it calls for knowledge-based aid and the globalization of knowledge requiring development organizations to reflect on how their internal Knowledge Management and learning systems interact with external information flows and policy trends (King, 2001).

In Ethiopia, development aid organizations play a crucial role in accelerating development assistance. The critical roles of these organizations are coordination of aid efforts, supporting development planning, building development capacity, investing in infrastructure and humanitarian aid.

Where knowledge sharing is concerned, the working culture of most of these organizations in Ethiopia is in a way that the focus is on practical questions of day-to-day work rather than on lessons being learned at a strategic level. It's usually hard to find good examples of a coordinated and strategic approach either to knowledge dissemination or knowledge strategy in the non-profit industry. The information services in these offices are mainly supply oriented, and user needs do not always coincide with the way information is made available.

Most of the time, the problems associated in this regard are not what technological platform to use but how to manage the vast amount of knowledge generated in a way that satisfies stakeholders, peers and government counterparts as a strategic asset for development assistance.

Although some development organizations try to actively align their work processes with a knowledge management component, these segregated tools (intranets, shared network drives etc.), case studies and best practices are not always available to other similar organizations. They merely serve as knowledge repositories for the specific organization's internal and usually mandate specific purposes.

The above stated challenges are main causes for the lack of known solutions to have a unified knowledge management practices where management, program staff, technical working groups synthesize their knowledge to reach development aid objectives.

Furthermore, there is an extensive and rich literature on knowledge management and its implications worldwide but most of these studies are conducted in the context of the profit making industries. There's little evidence, if any, regarding practical applications of knowledge management in development aid organizations in Ethiopia. It's critical to explore and implement knowledge management's relevance and impact on aid effectiveness in a development aid organization (Ramalingam, 2005).

It is essential that non-profit organizations working in the country have a viable strategy to manage and share their knowledge as well as learn from themselves. Without such a mechanism:

- Much knowledge must be re-learned in each region, project or programme, wasting costly time and resources
- Aid Organizations do not learn what is working and what needs to be changed in order to make their aid efforts more effective in Ethiopia.

To alleviate this, it is mandatory that an effective knowledge management initiative is in place which can be measured for effectiveness when considering enhancing the efficiency of development assistance. In order to establish an effective measurement of the impact of knowledge management we must first evaluate the structure of the knowledge flow and make up of knowledge sharing practices in the development aid organization in the country.

In light of the above, the primary objective of this research is to asses Knowledge Management maturity of development aid organizations in Ethiopia, and as a result to provide an insight for the development of a concise baseline. To achieve this, the study explored the following research questions:

- How do development aid organizations in Ethiopia use ICT and information management as enablers of knowledge management?
- Do development aid organizations in Ethiopia put in place strategies and policies to manage knowledge resources and/or to leverage existing knowledge?
- To what extent are development aid organizations in Ethiopia aware of the importance of knowledge management as a strategic enabler?
- To what extent do development aid organizations in Ethiopia rely on external knowledge sources to carry out their development assistance objectives?

The study assessed the level of knowledge sharing activities and suitable arrangements in place to facilitate the implementation of knowledge management initiatives in their respective organizations. Majority of the respondents indicated that their organization has strong knowledge sharing mechanisms even though not all mechanisms are yet systematic and well organized. Almost all respondents (90%) felt they have adequate ICT infrastructure in place that could support current or future KM initiatives.

The majority of the organizations who participated in this study recognize the importance of managing knowledge and claim to be making serious attempts to establish active KM initiatives. Respondents felt varying degrees of current and future KM initiatives represent a significant opportunity to improve their current work performance. Responses to a question on key challenges impeding effective knowledge sharing fell in the realm of lack of time or resources, some staff members being reluctant to share knowledge, concern that sensitive/confidential information becoming public and most importantly lack of formal organizational guidelines on proper knowledge sharing and management.

While the overall lack of well-defined KM strategies is a challenge, there are good practices and respondents have shared that their organizations are rapidly moving towards building on and further developing their knowledge management practices in the coming few years. Based on the responses gathered the study concludes by recommending action points that enhance conducive conditions for effective knowledge sharing in support of development goals.

Significance of the study

This study aimed to explore factors that influence the exchange of knowledge in aid and development organizations. The finding of this study demonstrated ways of putting in place effective knowledge management strategy and implementation in the development aid sector in Ethiopia. Output from this study will also strengthen sharing successful KM initiatives, learnt from the staff experience, both successes and failures, emerging from survey responses of experts.

The result from this research is expected to help in:-

- Assessing knowledge resources of all actors and concerned stakeholders in a way that guides reaching development aid objectives.
- Identification of information/knowledge needs, which will be compared with knowledge assets, followed by preparation of respective action plan.
- Expanding ways in which information and knowledge are disseminated, using explicit and tacit knowledge to solve practical problems.
- Capturing and sharing best practices and lessons learnt.

- Systematically include KM in development programmes of aid agencies in the country
- Give way to a development aid Knowledge Management Strategy, Streamlining management practices by institutionalizing KM or blending systematically KM processes with normal work processes.
- Strengthen government oriented capacity building operations.

Methodology followed

This study followed qualitative research methods via an online survey in the belief that this will achieve greater understanding and validation of results. This method was chosen due to the reason that the nature of the research problem calls for a description and identification of factors that portray existing conditions.

The research targeted development aid organizations in Ethiopia that have planned, implemented, or evaluated (to a certain extent) Knowledge Management related work processes and activities. The selected development aid organizations were identified on the basis of perceived relevance and access, giving priority to those that have relatively better internet availability, ICT infrastructure and relevant information management and knowledge management practices. They were selected based on the above criteria indicated either through professional partnerships, personal interactions and online professional communities with interests relevant to Knowledge Management.

In light of the above, the researcher decided to use purposive sampling for this research to be able to focus on the most relevant respondents for this study based on the principle that the selected respondents would most likely be significantly and directly interested in and/or involved in the phenomenon under investigation and hence provide a well-informed response. Therefore the sampling technique was selected to provide breadth of coverage.

The following organizations and country offices participated in the survey

- Action Aid Ethiopia
- Adoption Advocates International
- DKT Ethiopia
- Food for the Hungry
- GIZ Energy Coordination Office

- GOAL
- ILO
- International Food Policy Research Institute
- International Institution for Communication and Development
- International Livestock Research Institute (ILRI)
- International Rescue Committee
- Jhpiego
- Pact International
- Save the Children – Ethiopia
- United Nations Economic Commission for Africa
- UNICEF
- United Nations Office for the Africa Union
- United Nations Population Fund
- United Nations World Food Programme
- USAID
- Water Aid Ethiopia
- World Bank
- World Learning Ethiopia
- •World Vision Ethiopia

Data collection

Using quantitative research methodology, the study focused on empirical answers from knowledge management practitioners in the context of their organization's knowledge sharing practices. Structured data input were collected from the questionnaire that were distributed for this purpose.

After examining various research and practice-oriented publications including the Capability Maturity Model of the Software Engineering Institute (SEI) that many writers refer to, the researcher found that most knowledge management maturity models in the literature are ad hoc, and have not been empirically tested (Ehmset al, 2002; Harigopal et al, 2001)

To compensate for this shortfall, a knowledge management maturity measurement instrument developed by Kruger and Synman (2007) was used. The rational in using this survey instruments is two folds. 1) The researchers (Kruger and Synman) adhered to a research design that adequately combined theoretical propositions of knowledge management with practical applications while designing the knowledge management maturity questionnaire. 2) The questionnaires designed were not only benchmarked against known maturity assessment survey instruments, but also were thoroughly pre-tested and validated in collaboration with a team of knowledge management experts.

The above mentioned rational ensures that the survey instrument put in place can validate this research's empirical inquiry to rate factors that influence the success of Knowledge Management activities in development aid organizations in Ethiopia.

The online survey principally consisted of questionnaires specifically designed to measure knowledge management maturity. The instrument is composed of 101 descriptive questions under six sections namely;-

- ICT as an enabler of Knowledge Management
- Information Management in organizations
- Formulation of Knowledge Management principles, policy and strategy
- Implementation of Knowledge Management
- Ubiquitous knowledge transfer
- Assessment of Knowledge Management Growth

Each question was measured against a four-point scale with labels of Yes definitely, Yes but not significantly, No but probably within the next 5 years and No respectively.

Data were collected using primarily online survey questionnaire. The online tool Survey Monkey was used to collect survey responses. The questionnaire was sent to 30 selected individuals as a link via an e-mail containing all relevant information and expressing the strict confidentiality of their responses outside of the research interest. A mandatory consensual statement which verified that all participants are above 18 years of age was used to ensure that voluntary responses are based on adult judgment.

The on-line survey, based on Kruger and Synman's measurement instrument, queried many different aspects of capabilities for knowledge-sharing and management, which are summarized and discussed as follows.

A total of 24 organizations responded to this survey from a selected 30. The 30 organizations identified are based on the sampling criteria discussed earlier. By using the online data collection tool, the questionnaire was distributed to the 30 targeted organizations of which 24 responded via their national and international staff members who are responsible for knowledge management and communication practices in their organizations. These organizations have implemented at least a working information and knowledge sharing strategies that primarily serve an estimated 2,000 to 3,000 staff members as well as development partners and stakeholders.

Prior to answering the questionnaire the respondents were provided with an outline detailing the purpose and nature of the study, what the study focuses on and how the findings will benefit the non-profit industry in the country. In addition, the fact that the questionnaire was distributed over the internet resulted in wider coverage of the target population placing a great deal of convenience to many respondents at ease to participate in the survey. Face to face meetings, telephone conversations and e-mail messages were used to encourage participants to respond to the survey in a timely manner.

Findings

Despite being based on a small number of targeted population, the researcher believes that this paper has demonstrated the value of applying assessment in relation to knowledge management. The survey comments served to confirm the researcher's belief that the Kruger &Synman KM maturity assessment instrument and the general knowledge management maturity model (G-KM-MM) by Pee & Kankanhalli provide a useful framework through which to assess knowledge management.

Through a technique similar to critical success factor analysis, the three lenses of inquiry – information communication technology, strategic input and implementation issues - provide a rational and visible mechanism through which to conduct a knowledge management maturity and thereby assess an organization's current capability in relation to KM. With further research and development it is envisaged that the Kruger &Synman KM maturity assessment instrument will prove to be a useful diagnostic tool for those development aid organizations currently engaged in, or considering embarking upon, a knowledge management initiative.

Looking at the relations between the two chosen models used for this study (Kruger &Synman assessment instrument and the G-KMMM), even though one is a model and the other an assessment instrument, the fundamental issues discussed by both authors have similarities. G-KMMM emphasizes maturity levels against key performance indicators of three core pillars: (people, process and technology). Each pillar is described by a set of characteristics that explain specify practices that, when collectively employed, can help organizations accomplish the goals of the particular maturity level they most likely resemble. The GMMK has five maturity levels that are measured against the key performance indicators namely, initial:- where little or no intention of KM is exhibited, aware:- Organization is aware of and has the intention to manage its organizational knowledge, but it might not know how to do so, defined:- basic infrastructure is put in place to support KM, managed:- KM initiatives are well established in the organization and optimized:- where KM is adequately integrated into organizational processes.

While Kruger &Synman focus on 6 main areas of assessment when evaluating KM in an organization. Kruger and Sunman's model incorporates the three main pillars of the G-KMMM in that the first two critical success factors:- ICT as an enabler of Knowledge Management and Information Management's role for knowledge management in an organization describe the technological readiness of the organization. Similarly the two other success factors of Kruger and Synman namely Formulation of Knowledge Management principles, policy and strategy in an organization and 'Implementation of Knowledge Management in an organization' touch upon the processes involved for the smooth implementation of KM. The two factors assess the organizational processes that are essential for establishing a successful knowledge management initiative.

Kruger &Synman instrument also emphasizes on ubiquitous knowledge transfer which elaborates on the soft issues of KM implementation that primarily involve people's perception of KM processes and their behaviours towards it which directly relates to the GMMK's core pillar.

Based on the exploratory survey conducted, it appears that the vast majority of the firms examined are doing at least some initiatives in the name of either Information Management and/or Knowledge Management. It was found that higher levels of organization readiness for KM were indicative of more commitment and less pessimism about KM.

The majority of the organizations who participated in this study recognize the importance of managing knowledge and claim to be making serious attempts to establish active KM initiatives.

Despite this enthusiasm, the actual implementation of KM still tends to be conducted on an informal, ad hoc basis with little evidence of effective co-ordination among departments.

The major findings of this survey can be summarized as: -
- There's a rapidly growing interest and engagement to develop an effective KM initiative as an enabler of development assistance objectives within the non for profit sector in the country.
- Most organizations possess adequate ICT infrastructure that facilitate the implementation and growth of knowledge management initiatives.

- Most organizations have already in place a working knowledge management and information sharing systems even though a lot of them don't yet follow a systematic, well organized mechanisms of retaining and disseminating knowledge.
- Staff and knowledge workers in these organizations are actively involved in sharing information and knowledge resources as and when required for speeding up working processes. This positive attitude towards knowledge sharing with colleagues and stakeholders is a key conducive behaviour to boost growth of KM.
- Albeit incomplete, development aid organization have put in place at least the basic prerequisites and strategic arrangements that enhance the sharing of knowledge within and outside their organizations.
- KM implementation challenges are well defined. The majority of the respondents find the absence of proper organizational guidelines on knowledge sharing, lack of knowledge of what colleagues need, and shortage of time and resources to facilitate knowledge sharing hindering their desire to share knowledge with colleagues within and outside the organization.
- Development aid organizations are actively involved in social mobilizations and group learning activities with their stakeholders through various communities of practice that frequently interact through face to face meetings.
- Despite all the above mentioned achievements, devising and implementing KM strategies aligned with the organizations' strategic priorities to guide future KM directions is still at a very early stage.
- Matching analysis of the results of the study against the G-KMMM model, most organizations fall under the third maturity level of the model titled 'Defined' (see Table 2).

While many of the respondents appeared to be critical of their respective organizations' development of KM strategy, they did feel that KM initiatives, current and future, represented a significant opportunity to improve their current work performance. Issues such as lack of time or resources, some staff members being reluctant to share knowledge and concern that sensitive/confidential information becoming public were mentioned as major hindrances to active knowledge sharing practices.

Recommendations

Like most development assistance programs around the world, international NGOs in Ethiopia face the need to improve the content, process and inclusivity of impact of their development assistance efforts. Most of these organization are knowledge based, processing data and key information for decision making on a daily basis. It is therefore pertinent that these organizations are capable of increasing knowledge in the development sectors they operate and/or leverage existing knowledge to meet development goals. This calls for an efficient knowledge management strategies coupled with acquiring the necessary leadership and support for implementation.

Looking at the diffusion of KM initiatives throughout an organization using a change lens, the process of implementing KM would be expected to unfold through a series of stages. Readiness, the initial stage, would occur when the organizational members' attitudes are such that they are receptive to a forthcoming KM effort. Adoption occurs when the organizational members alter their attitudes and behaviours to conform to the expectations of the KM effort. Institutionalization occurs when KM becomes a stable part of employees' behaviour and fabric of the organization. According to the responses, most organizations are at the readiness level.

Because knowledge management implementation and integration is a cultural change process, which involves changing the hearts and minds and work habits of staff, it can't be done overnight. It may take at least a couple of years before KM is fully institutionalized across the entire organization. To achieve that development aid organizations need to continue their dedicated implementation efforts, focused on breakthrough and delivery of new ways of working, which lasts until they start to establish rules and disciplines for KM, a knowledge management strategy.

Based on the survey results and considering the observed KM readiness of the organizations who participated in this study, the author forwards the following recommendations that further enhance conducive conditions for effective knowledge sharing that support development goals and in the long run can integrate leveraging organizational knowledge within strategic directions.

- Initiating formal trainings (including e-learning based), peer learning, experience sharing and similar other formal and informal learning interactions, through knowledge management initiatives, to meet the knowledge needs of the majority of staff.

- Provision of regular training and practical opportunities aimed at building and sustaining the different knowledge creation and sharing skills to

all staff of different levels. Establishing knowledge sharing skills development strategies as well as programs to popularize knowledge sharing among the staff of the organizations.
- Strengthening peer learning schemes using different mechanisms (such as communities of practices and regular e-discussion) that best allow bringing together the knowledge of staff members with experts from outside of the organization to ensure that continuous and focused learning takes places within and outside development aid organizations.
- Virtual communities of practices could also be used for facilitating exchange of knowledge across units. Initiating discussion forums, composed of face-to-face and e-discussion, to create the platform that would enables sharing experiences and lessons in turn building an environment of trust with all its stakeholders and partners on a regular basis.
- Building knowledge sharing environment by combining the already popular knowledge sharing mechanisms (such as face-to-face meetings and e-mail) with those not widely used e by the respondents (such as e-discussions and virtual CoPs) in a balanced way to make the most of available knowledge sharing mechanisms.
- Encouraging the use of e-mail based e-discussions. The organizations could then build on the skills, experience, and interest that staff would have acquired in using the e-mail based e-discussions to introduce and popularize web-based e-discussions and virtual forums.
- Making accessible knowledge resources (both tacit and explicit) to encourage self-teaching (including e-learning resources) particularly on those areas of knowledge identified as strategic
- input. Training and resource building by the organization therefore should be built around meeting the knowledge needs of the staff.
- Making sources of knowledge relevant to the development objectives of the organization more accessible both physically and electronically visible to encourage more use of the sources. Making accessible knowledge products such as lessons learned documents, best practices documents, peer learn files, and after action reports to all stakeholders involved in common areas of concern are one of the key strategies that the knowledge management program has to consider to strengthen discussions and exchange of knowledge across units of the organization.
- Making knowledge products easily accessible from centralized repository of easy to use interface that can be accessed from different locations

as prioritized by the knowledge needs and other relevant criteria that the development assistance programs require.
- Strengthening and financing units responsible for knowledge management to facilitate further identification of knowledge needs, training needs, barriers to knowledge sharing, and so on.
- Although the positive attitude towards sharing knowledge is an important asset for development aid organizations in the country, more work needs to be done by the organizations to develop the appreciation and enthusiasm in all staff to engage knowledge sharing with experts and stakeholders outside of the organization as well. It is important for the organizations to instil the value of the importance of knowledge sharing across the board to make knowledge sharing a norm. Creating awareness as to the personal and organizational values of knowledge and knowledge sharing, provision of incentives for engaging in organizational knowledge sharing (through institutional recognition of time spend on knowledge sharing, making knowledge sharing as one area of competence the performance of staff would be assessed for, and so on), and provision of tools and skill sets required to engage in knowledge sharing are the primary measures that the organization has to take in order to improve knowledge sharing.
- Putting in place proper organizational guidelines, proper tools, and identification of the knowledge that need to be shared should be considered as one of the primary measures to be taken by the organizations to build their knowledge management programs on solid foundations. This can be addressed through developing knowledge management strategy as one of the first steps.
- The researcher hopes this investigation serves as a building block for subsequent research that not only explores other facets of KM maturity assessment in development aid organizations, but also advances the level of sophistication of the research.
- Future research efforts can use this work to guide studies that examine the entire process that organizations might go through as KM is introduced. It is envisaged that future studies, based on these findings, will build on providing a pre and post implementation assessments of the success of KM initiatives, hopefully with wider scope and coverage. Furthermore, an in-depth research is needed to evaluate the direct or indirect relation between the knowledge management practices already in use and the implementation effectiveness of development assistance objectives.

REFERENCES

Maciocha, A., and Kisielnicki, J., (2008). Intellectual Capital and Corporate Performance. *The Electronic Journal of Knowledge Management.* 9(3), 271-283.

Fowler, A., . (1992). Prioritizing Institutional Development: A New Role for NGO Centres for Study and Development. International Institute for Environment and Development. UK.

Maizels, A., and Nissanke, M. K., (1984). Motivations for aid to developing countries. USA, Elsevier Inc.

Andersen, A. (1996) The Knowledge Management Tool. Developed jointly by Arthur Andersen & The American Productivity and Quality Centre, Chicago, USA.

Alavi, M., Leidner, D. E. (2001).Review: Knowledge Management and Knowledge Management Systems: Conceptual Foundations and Research Issues. *MIS Quarterly.* 25(1), 107-136.

Ramalingam, B., (2005). Implementing Knowledge Strategies: Lessons from international development agencies: Overseas Development Institute (ODI). UK

Boisot, M. (1987). Information and Organizations: The Manager as Anthropologist. Fontana, Collins.

Macneil, C., . (2001). The Supervisor aa Facilitator of Informal Learning in Work Teams. *Journal of Workplace Learning,* 13(6), 246-253.

Davenport, T. H., and Prusak, L., 1998. Working Knowledge: USA, HBS Press.

Hipshop, D., (2003). Linking Human Resource Management and Knowledge Management via Commitment: A review and research agenda. *Employee Relations,* 25(2).

Szczerbicki, E., and Nguyen, N. T., (2010). Smart Information and Knowledge Management: USA, Springer.

Leborne, E., and Hearn, S., (2011). Monitoring and evaluating development as a knowledge industry: ideas in current practice: IKM Emergent. Working Paper No. 12, GERMANY.

Feagin, J.R, Orum A.M & Sjoberg G. (1991).A Case for the Case Study. USA, UNC Press.

Frid, R. (2003). A Common KM Framework For The Government of Canada: Framework For Enterprise Knowledge Management. Canadian Institute of Knowledge Management. CANADA.

Gashaw Kebede. (2010). KM an information science perspective: *International Journal of Information Management.* 30, 416-424.

Georg Hüttenegger. (2003). Knowledge Management System Building Blocks: *EJKM,* 1(2), 137-148.

Hedlund, G. and Nonaka, I. (1993). Models of Knowledge Management in the West and Japan. USA, Macmillan.

Henderson, K. (2005). The knowledge sharing approach of the United Nations Development Programme :*KM4D Journal.*1(2), 19-30.

IFAD. (2007). Knowledge Management Strategy. Palombi e Lanci. ITALY.

Irma Becerra-Fernandez, Rajiv Sabherwal. (2010). Knowledge Management Systems and Processes. USA, M.E. Sharpe, Inc.

Dalkir, K., (2007). KM in theory and practice.UK, Elsevier Butterworth-Heinemann.

Kogut, B. & Zander, U. (1993). Knowledge of the Firm and the Evolutionary: Theory of the Multinational Corporation. Journal of International Business Studies, 24, 625–645.

Demarest, M., (1997). Knowledge Management: An Introduction. USA, Elsevier.

McElroy, M.W.(2000). Integrating Complexity Theory, knowledge management and organizational learning: *Journal of knowledge management.* 4(3), 195-203.

Mthuli Ncube, Charles Leyeka Lufumpa, Leonce Ndikumana (2010), Ethiopia's Economic Growth Performance: Current Situation and Challenges, *Economic Brief,* 1(5), 1-5.

Nonaka, I. (1994). A Dynamic Theory of Organizational Knowledge Creation. Organization Science. 5(1), 14-37.

Organization for Economic Cooperation and Development. (2008). DAC Report on Multilateral Aid. SWITZERLAND.

Griffiths, P., and Remenyi, D., .(2008). Aligning Knowledge Management with Competitive Strategy: A Framework. *Electronic Journal of Knowledge Management,* 6(2), 125 – 134.

Mosley, P. (1980). Aid, savings and growth revisited. Oxford Bulletin of Economics and Statistics. 42(2), 79-95.

Drucker, P., . (1994). Post-Capitalist Society. USA, HarperCollins Publishers.

McKinlay, R. D., and Little, R., (1979). The U.S Aid Relationship: A Test of the Recipient Need and the Donor Interest Models. *Political Studies*. 27(2), 236-250.

Callahan, S,. (2009). Crafting a Knowledge strategy: USA, Anecdotepty Ltd.

Stankosky and Baldanza. (2001). A Systems Approach To Engineering A KM System. [unpublished manuscript]

Scwandt, T.A.(1994). Constructive, interpretivist approaches to human inquiry. Sage Publications.

Mansfield, M., and Grunewald, P., (2013). The use of indicators for the Monitoring and Evaluation of knowledge Management and Knowledge Brokering in International Development.UK, Loughborough University.

Yin, R.K.(1984). Case Study Research: Design and Methods. USA, Sage Publications.

CHAPTER 18:
A Learning Oriented Approach for Organizational Requirements Development

Mesfin Kifle

INTRODUCTION

The dynamism and complexity of organizations is increasing as time goes on. Deciding on features of a new software system is becoming challenging to realize the required behaviour in the environment it embeds. Nevertheless, many organizations in developing nations like Ethiopia are attempting to use ICT to leapfrog their development and use software as a means for development and being competitive.

In organizational system development, an organization is considered as the domain where the software is applied in the context of human activity to change the existing situation. But, organizational situation is by far complex. To specify what the new system should be, it is advisable to understand different aspects from different viewpoints to realize the required behaviour.

Requirements development (RD), as part of system development in context, refers to the overall process of understanding the as-is situation of an organization, what is needed to improve, what are the to-be requirements, and how do we approach to do these. Requirements indicate what users need from the system and are described in terms of their effect in the environment.

The notion of requirements in an organization refers to business requirements, user requirements, product requirements, project requirements, etc.(Weiger, 2003)(Wiegers, 2000). According to the reference model of requirements and specification proposed by Gunter et al. (Gunter, Gunter, Labs, Jackson, & Zave, 1998)"given" : "Michael", "non-dropping-particle" : "", "parse-names" : false, "suffix" : "" }, { "dropping-particle" : "", "family" : "Zave", "given" : "Pamela", "non-dropping-particle" : "", "parse-names" : false, "suffix" : "" }], "id" : "ITEM-1", "issued" : { "date-parts" : [["1998"]] }, "title" : "A Reference Model for Requirements and Speci cations University of Pennsylvania", "type" : "article-journal" }, "uris" : ["http://www.mendeley.com/documents/?uuid=b9bf543c-d4bd-4b0c-90e0-f97b-8c7245df"] }], "mendeley" : { "previouslyFormattedCitation" : "(Gunter, Gunter, Labs, Jackson, & Zave, 1998, requirements can be described from general system perspective in relation to domain and environmental properties.

But, in general, it is hardly possible to understand and manage organizational requirements without continuous and step-by-step communication with users. Situational and contextual factors influence stakeholders' decision. Such factors are multi-dimensional in their origin and construction. Some of the issues might be observed as organizational behaviours that possibly emerge in the history of the organization, social and cultural status, economic status, believes, etc. Furthermore, other factors such as individual experience, nature of work processes and practices might also become possible causes to determine contextual settings. These occur in a specific or across similar organizational environment (Cornut, 2009).

Thus, in reflection to these factors, it is strongly recommended to assess 'suitable' learning oriented approaches for requirements development that enable active users' participation, promote organizational learning and enlighten with goal orientation.

METHODOLOGICAL APPROACHES IN CONTEXT

In requirements development, it is a common practice to deal with purpose and context of an organization to make a system *fit-to-purpose* and *fit-to-context*. In Soft System Methodology (SSM), the human factor is discussed in line with the notion of problematical situation of an organization (P. Checkland and J. Scholes, 1990). Furthermore, since the situation in an organization is complex, understanding the problematical situation is important before taking action for change. Therefore, in modelling, it is essential to address issues that concern people in a way that represent their viewpoint. It is also required to get root definition of a system to proceed.

Furthermore, according to Reflective System Development (Mathiassen, 1998) problem-setting is a process and if developers fail to engage in problem setting, they may be in a position either to misunderstand the situation or limit their understanding. Hence, to understand, support, and improve systems development, developers must go beyond theories and methods and understand the role of knowledge and the relation between knowledge and action in systems development practice.

The Reflective Steps approach, proposed by Biru (T. Biru, 2008), has given more emphasis on transformational type of participation taking experience based learning oriented approach through reflection. He underlined that collaborative reflection is important in designing a product and a process in a given context. It is also discussed that success of collaborative reflective learning depends on factors related to participants' knowledge, group combination, individual participant's outlook and contribution.

Application of collective mind theory in requirements development is elaborated by Crowston et al. (K. Crowston and E. E. Kammerer, 1998) and Hsu et al. (J. Hsu, T. Liang, G. Klein, 2010). Individual and group level learning, understanding and building performance have been suggested to elaborate users' contribution in the process.

Therefore, an approach that fosters user-developer collaboration to build shared understanding about the existing and anticipated (to-be) organizational knowledge is paramount. Understanding of users' situation can lead to more innovations. Defining the problem setting in organizations particularly in developing nations is challenging due to the reasons including:-there are multiple views and knowledge gaps in organizations, there are limitations and gaps on aligning IT systems to support activities, processes and goals, and system development activities lack to anticipate and pursue changes effectively in line with organizational goals.

Along with the facts raised above regarding to organizational requirements development, the research questions identified for investigation were: what are the "practices" in defining organizational requirements that have been used across requirements development projects in academic information management systems and how a learning oriented approach contributes to address the situation.

METHODOLOGY

In order to get better understanding of the situation and practice of requirements development at the public universities in Ethiopia and answer the research questions, it was required to conduct an empirical investigation. Due to the researcher's past experience, ease accessibility and having long history, Addis Ababa University (AAU) has been taken as a case to conduct in depth study. To this end, a qualitative research approach has been preferred.

The methods selected to use were survey and case study. Both were used to do situational analysis. The purpose of the investigation was to study patterns and variability factors of employee task knowledge at AAU and organizational requirements development practices at six public higher learning institutions in Ethiopia, namely, Jimma, Mekele, Bahr Dar, Adama, Gonder and Addis Ababa University so as to find out situational factors that impair or facilitate requirements development. These six universities have been selected since they are established long before and have relatively better experience in requirements development. Therefore, it is assumed that the aforementioned universities can sufficiently represent the domain for this research.

A survey is used to get a more apparent picture of the requirements development practice. Assessments were made to understand the situation through 1) open-ended interviews with the main responsible actors in software development projects and 2) review of requirements documents. Analysis of data is done with themes corresponding with the issues – user participation, domain and goal orientation. The survey was conducted about aspects of existing campus management system projects such as their failure & success, challenges in requirements development, etc. at the six public universities.

Case study is among the common qualitative research approaches that are used to study a given phenomenon in depth taking specific events or projects as cases. It is used for the purpose of practice studies – situational analysis. Case studies are descriptive reports of episodes. Regarding the strength and weaknesses of cases studies from a research perspective, Cockburn (Cockburn, 2003) wrote, "each case study provides a data point. The strength of the case study is that it captures the local situation in greater detail and with respect to more variables than is possible with surveys. ... Case studies are helpful in developing and refining generalizable concepts and frames of reference..." (p. 29).

Case studies can also be used to study complex phenomena in its institutional context. Yin (Yin, 1994) defines case study "as an empirical inquiry that investigates a contemporary phenomenon within its real-life context..." (p. 13).

In this research context, the purpose of the case study was to explore organizational, human and technical factors that have impact on requirements development. It was done at the office of the registrar, AAU.

FINDINGS OF SITUATIONAL ANALYSIS

While attempting to identify key issues, in addition to the researcher's experience in system development and working at AAU and a utility company, similar attempts have been reviewed including the work of (R. Kifle, 2004) and (T. Biru, 2008). Both attempts were made to assess the practice in software development in general. The findings related to requirements development in the two studies have been noted down to find the gap and do further survey. In this research, to deal with relevant issues in depth and enrich the previous findings, emphasis was given for requirements development in view of user participation, goal orientation, and requirements development approaches. For this purpose, structured open-ended qualitative interview type was preferred to design.

Once the initial design of the interview was drafted, it was made to try out the interview with three experts at AAU. All the experts have masters' degree in ICT related areas and by the time when the interview conducted, they were serving in three different positions: as system administrator of the registrar system, as project manager of the budget and financial system project, and application development unit head of the University ICT office. After reviewing the interview results, the issues were further refined before moving to the next steps.

In the selection of interviewee in each institution, the researcher was guided by the roles individuals play according to the organizational structure. All the selected institutions had ICT office which is fully responsible in software development within the institution. The researcher attempted to reach to all individuals in the office who had stake in a particular software development project. Though it was planned to conduct the interview one-to-one, whenever possible and the situation allowed it was also conducted in group.

In general, a total of 18 ICT professionals in the six public universities were purposefully selected in the survey. In the time range from Oct 2011 to Dec 2012, the researcher physically travelled into the five universities outside Addis Ababa and conducted the interview at their premises. Each interview session lasted on the average for 45 minutes. The interview and discussions were conducted in the Amharic language. The interview and discussion points which were noted down were then consolidated and interpreted in view of the issues identified in this research. Moreover, the interviewee provided relevant requirements documents which had been used in a particular project. Not all documents are the same in terms of content and completeness. While some are very brief description of system requirements others include a detailed description of the requirements anticipated from the system-to-be. These documents were reviewed by the researcher.

The number of interviewee and projects assessed in the survey is shown in table 1.

Table 1: Summary of the interviewee involved and projects assessed in the survey

Public University Name	No of Interviewee	No of Software Development Projects Assessed
Addis Ababa University	3	2
Adama Science and Technology University	4	5
Bahir Dar University	2	2
Jimma University	2	3
Mekele University	4	2
University of Gonder	3	2
Total	18	16

A total of 16 projects were assessed in the survey. Among these, 8 of them were in-house projects with an attempt to customize open-source products. Among these, only 3 were partly implemented in the institutions' business environment and they were small projects. It is observed that in relatively large projects, problems in requirements development were complex. The remaining eight projects were either outsourced or customization of commercially available packages or open-source products.

Due to the initiative of the government to reengineer public services across the country, since 2008, all the six universities included in the survey have conducted business process reengineering. As a result, the business context was continuously changing and 50% of the projects were affected. Moreover, the ICT policy of each institution had impact on the way projects were designed and managed. While three of the institutions were more inclined to adopt open source principles and policies, the other three did not have a strong stand on this matter and heavily relied on outsourced projects. In all these projects, only the conventional requirements development methods including document review, interview, and observations have been used to understand and describe the requirements.

According to the survey results, the level of user participation was minimal in all the projects assessed. Lack of a means to share knowledge with users is a common problem and the problem was aggravated due to lack of a common language. Similar problem was also noticed in the survey conducted by (T. Biru, 2008).

The situation was worse in one institution which had a strategy to customize and use open source products. Users' participation was little and the IT professionals were not experienced and trained in methods that promote user participation.

Another important factor that hindered user participation is that users lose their motivation as the project progresses though they are eager to see positive impacts of the to-be systems immediately on their working environment. But, only 50% of the projects have been completed partially at the time of project completion.

Therefore, in order to increase quality, it is worth noting to address active user-participation particularly at the time of organizational requirements development to increase understanding of organizational context and react accordingly.

Regarding goal orientation, it was found out that all professionals at least believed that software development projects should be aligned and support business goals and processes set out for institutions. But, in practice, the attempt made to define projects in line with business goals taking possible opportunities and traits into account was not satisfactory in most of the institutions. It was also noticed that the existing environment does not allow realizing some of the business goals anticipated in the BPR process. Even, some have faced challenges to set a strategy that directs to the intended goals. On the other hand, top decision makers were not active to learn the situation and respond promptly.

The attempt in one of the institution was relatively better than the others. In the remaining five, though the business goals have been used to drive the requirements development process, less is done or with limited capability in aligning product requirements with goals. .

Moreover, it has been noticed that as requirement experts able to understand and use organizational goals in requirements development, communication and collaboration with users becomes better.

Findings of Employee Task Audit- the Case Study

The case study is conducted at the office of the registrar, AAU, to explore and study employee knowledge about their day to day tasks. The following three major goals were set out for the study: understand employees' knowledge and detect pattern of tasks and variability that can possibly be exhibited, build shared understanding among employees across the different campuses of AAU and finally explore the use context of the existing registrar system.

The study was conducted by the researcher collaboratively with 11 campus registrar unit heads and one IT section head of the office. The researcher acted as a leader and involved on activities including planning, preparing task audit samples by investigating the registrar system database, assign task audit assignment for each head, review audit reports, organize and facilitate face-to-face plenary discussions, reflect on the actions taken, prepare for action to improve 'bad' situations on some campus registrar units.

During the study, four task accomplishment or performance audits have been conducted including employees' experience in the existing registrar system. And then, based on the audit reports, consecutive plenary face-to-face discussions were made with heads of the units. During these meetings, the unit heads were prompted to reflect on 'what was intended to achieve', 'what had been observed during the study' and 'what can be done to improve.' Due process, the heads were able to share knowledge, understand the strengths and weakness of the existing system, and suggest alternative solutions to improve the situation. They were also able to detect and understand knowledge variability.

Consequently, based on the analysis result across the campuses, variability factors that affect organizational requirements were identified in relation to employees' knowledge about their tasks. These factors are:

- *Location factor:* in the study it was observed that employees in seven registrar units had developed their own knowledge, which was unique to each unit. The two basic reasons which have contributed for this are: the unique type of problems that had happened in the units and prior experience of the units. For instance two campuses were independent institutes before they were integrated to the University, and had their own practices.
- *Experience and competency factor:* due to self-competency and range

of work experience some employees have developed their own set of activities to accomplish a similar task than others. Such situations have also been observed in the same location.

- *Education type factor*: employee educational qualification type had also affects the nature of employee knowledge while doing their job. For example, for roles like registrar unit heads, the educational qualification requirement is basically defined based on educational level than type. To be a unit head, having a bachelor degree in any field with some years of experience was enough.

Taken together, the lesson taken from these findings point out that, in order to have a wider and in depth understanding about the working environment and situational problems in organizations, it is necessary to have a set of criteria during organizational requirements development so as to reach to the various types of employee knowledge, detect variability and understand what has to be done. Attention should also be given to domain knowledge and business goals.

A LEARNING OREIENTED APPROACH

The main aspects of the proposed ontology based learning oriented approach are shown in figure 1.

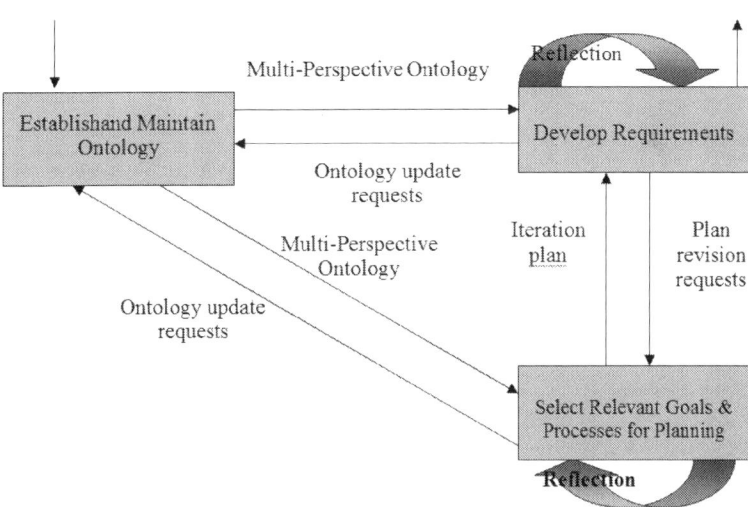

Figure 1: High-Level Block Diagram of the proposed approach

As can be seen, it identifies three major integrated activities. Namely: *Establish and maintain ontology, Select Relevant Goals and Processes for planning, and Develop Requirements.*

Establish and Maintain Ontology is an activity introduced in the proposed approach by reflecting on the need to capture, maintain and use the required concepts as per defined in the organizational ontology model described in (M. Kifle, 2012). It helps to continually and iteratively support organizational learning. Since understanding an organizational context is a result of continuous learning process, the ontology content evolves over a project time until the requirements development is reached to the desired level. Indeed, the *ontology update request* arrow is shown in the diagram to reflect this fact in relation to the other two activities.

At initial stage of a software development project a requirement expert shall conduct preliminary assessment about the domain of interest and scope of the project to have first-hand impression and high level understanding of the project. Techniques like document review, interview and any other which are deemed appropriate can be used for the purpose and start to maintain the ontology. Domain experts within the organization can also assist on updating the ontology.

Select Relevant Goals and Processes for planning is the second activity introduced in the proposed approach by reflecting on the need to learn and understand contextual issues iteratively and incrementally. It is a means to deal matters like user group establishment and define a reflection process. In each round, an *iteration plan* is prepared to guide the third activity, *develop requirements*.

Develop Requirements is the third activity introduced in the proposed approach in which the product requirements development is carried out. It is a process to understand the *as-is situation, envision the to-be* system and define requirements. The learning oriented process is mainly carried out in this activity. Requirements definition is the final output of the activity. An ontology update request and iteration plan revision might be initiated depending on the learning curve users and developers will be.

The proposed learning oriented approach was applied in a small scale on the requirements development project at AAU in 2012. The results obtained are encouraging. It enables to explore and identify requirements which were overlooked in the previous attempts.

CONCLUSION

In this paper, it is reflected that an organization is considered as the domain where application software is applied in the context of human activity to change the existing situation. The situation in large organizations is by far complex particularly in developing nations. Requirements describe what is needed from the new system and it is hardly possible to easily understand and manage organizational requirements without continuous and step-by-step communication with users. Understanding of the users' situation can lead to more innovations.

The two studies described in this paper have shown that during organizational requirements development attention should be given to learning oriented active user participation, domain orientation and goal-orientation. Moreover, taking the following factors into account is very important: *location factor, experience and competency factor, and education type factor.*

As a result, in this research, ontology based learning oriented approach is recommended. It is an iterative and incremental process in reflection to participants' level of understanding. Further work is needed to evaluate its effectiveness.

REFERENCES

Cockburn, A. (2003). *People and Methodologies in Software Development*. PhD Dissertation, University of Oslo, Norway.

Cornut, F. (2009). *The Discursive Constitution of Software Development*. PhD Dissertation, London School of Economics and Political Science.

Gunter, C. A., Gunter, E. L., Labs, B., Jackson, M., & Zave, P. (1998). A Reference Model for Requirements and Specifications, University of Pennsylvania.

J. Hsu, T. Liang, G. Klein, and J. J. (2010). Promoting the integration of users and developers to achieve a collective mind through the screening of information system projects. *International Journal of Project Management*, 1–11.

K. Crowston and E. E. Kammerer. (1998). Coordination and Collective Mind in Software Requirements Development. *IBM Systems Journal*, 37, 227–245.

Kifle, M. (2012). Multi-Perspective Ontology to Understand Organizational Requirements. *Proceedings of the African Conference on Software Engineering and Applied Computing, Gaborone, Conference Article, Botswana* (pp. 21–27).

Kifle, R. (2004). *Software Development Assessment in Ethiopia*. Masters Thesis, Department of Computer Science, Addis Ababa University.

Mathiassen, L. (1998). Reflective Systems Development, *Scandinavian Journal of Information Systems*, *10*(1 & 2), 67-117.

P. Checkland and J. Scholes. (1990). Soft Systems Methodology in Action. New York: John Wiley and Sons.

T. Biru. (2008). *Reflective Steps: A Collaborative Learning Oriented Approach to Software Development and Process Improvement. Forschung*. PhD Dissertation, Hamburg University.

Weiger, K. E. (2003). *Software Requirements* (2nd Ed.). Washington, USA, Microsoft Press..

Wiegers, K. E. (2000). Karl Wiegers Describes 10 Requirements Traps to Avoid. *Proceedings of Software Testing & Quality Engineering,* 2(1), 1 - 8.

Yin, R. (1994). *Case Study Research: Design and methods* (2nd ed.). Thousand Oaks, CA.Sage Publishing.

CHAPTER 19:
Understanding Community Resource-Use Linkages: A Case Study Conducted in Southeast and South-central Alaska

Mekbeb E. Tessema, Robert J. Lilieholm,

Dale J. Blahna, Linda E. Kruger,

and Joanna Endter-Wada

Introduction

Social views of the appropriate management and use of public forest lands in the U.S. have undergone significant changes over the last 40 years, evolveing from the proveson of market commodities like timber and grazing, to the protection and maintenance of ecosystem services (Kennedy and Koch 2004, Kruger and Mazza 2006). This transition has in part been driven by a host of laws, including the National Environmental Policy Act (1969), the Endangered Species Act (1973), the Resources Planning Act (1974), the National Forest Management Act (1976), and others. These laws – and their interpretations under the legal system – have profoundly altered the management of hundreds of millions of acres of public lands, perhaps none more so than the 191 million acres within the National Forest System (NFS) managed by the U.S. Forest Service (USFS).

Against this backdrop of changing mandates toward public land management, forest use in the State of Alaska presents an interesting case where nearly 89% of the land base is publicly owned, 10% is under Native corporate control, and just 1% is privately held. There, extractive uses – including mining, logging, firewood cutting, and hunting – support many rural Alaskan communities, especially Native communities, yet this historic connection between communities and public lands is challenged by national trends that shift management away from the production of commodities like timber, and toward greater emphasis on cultural values (e.g., spiritual, recreational, and aesthetic) and ecological services (e.g., regulating and supporting services like wildlife habitat and water quality) (Deal et al. 2012).

Nearly 38% of Alaskans live in rural communities with fewer than 10,000 residents. Moreover, nearly 17% of Alaskans are Alaska Natives, with 69% living in rural communities, most of which reside in the vicinity of the Tongass and Chugach National Forests (TNF and CNF, respectively) (Allen *et al.* 1998, USFS 2007). Alaska's many rural communities, combined with the dominance of public lands, makes local access to public resources a critical issue.

Historically, limited data have hampered the ability of agencies to understand these community-resource linkages and to use the information in planning and decision-making (Endter-Wada and Blahna 2011). This paper presents a practical method for using resource use permits issued by the USFS to help describe how communities are directly linked to natural resources in Southeast and Southcentral Alaska. The method also addresses other social assessment weaknesses such as difficulties identifying a zone of social influence for agency actions, and presenting data as a visual display that can used to compare and contrast with biophysical information.

Community Access to Public Land Resources

Access to natural resources is critical for the social, cultural and economic well-being of rural resource-dependent communities, and in many parts of the world, rural poverty and lack of access to natural resources are positively correlated (Cook 1995, Arnold and Townson 1998, Leach et al. 1999, Wood and Holling 2002). For public lands, communities may lack access for multiple reasons, ranging from direct physical exclusion, to land use regulations and other policies that reflect shifting perspectives on the value and appropriate use of resources (Carroll 1991, Lee 1991, Humphrey et al. 1993). Game laws, the establishment of parks and nature reserves, and the protection of endangered species are common examples (Ives 1988). Some impacts may be indirect – e.g., when resource protection increases land values and excludes local people from access (Geisler and Mitsuda 1987, Fitchen 1991).

Evaluating public land management policies in the context of local resource access is important to the social, economic, and cultural well-being of rural communities (Knutsson and Ostwald 2006). The relationship between poverty and lack of access in resource dependent communities has been addressed by multiple theories that focus on community characteristics. Stedman (1999) and Stedman et al. (2004) identified important factors to include: (1) lack of human capital in rural areas; (2) low income and benefits due to the "peripheral" nature of industrial structures (e.g., job seasonality in tourism sectors); (3) power and natural resource bureaucracy (i.e., extractive sectors may be owned by powerful interests, which may limit local access and decision-making); and (4) moral exclusion (e.g., popular sentiment may exclude some forms of resource extraction).

Understanding the nature of community-resource linkages in the context of the private use of public lands is critical to sustaining human and natural systems. These linkages can vary considerably, depending on the resources available, land management policies (Endter-Wada and Blahna 2011), and a community's cultural, demographic, and economic characteristics (Jakes et al. 1998). In Alaska, the issue of local access to public resources is particularly important because nearly 40% of residents are rural and often dependent to some extent upon publicly owned lands to provide various uses and services (Allen et al. 1998). Policies regarding access to resources may affect rural or Native communities' livelihoods, and also important cultural and subsistence practices.

Endter-Wada and Blahna (2011) examined U.S. public land and resource access and use policies and developed the "Linkages to Public Lands" (LPL) framework for describing human-resource linkages on public lands. The LPL framework identifies five types of community-public land linkages: tribal linkages, interest linkages, neighboring linkages, decision-making linkages, and use linkages. Whereas most community assessments used in resource planning describe characteristics of nearby communities and then assume that some generalized resource-use linkages must exist, LPL differs in that it first identifies the actual resource-use linkages and then identifies the communities where the people in those linkages reside.

While each type of LPL linkage is important, two are prominent in Alaska. First, tribal linkages to public lands are central to the existence and identity of many Native communities and are recognized by the 1971 Alaska Native Claims Settlement Act (ANCSA) and the 1980 Alaskan National Interest Lands Conservation Act (ANILCA). Under these laws, Alaskan tribes have access rights to USFS lands and resources, as well as a "government-to-government" relationship with state and federal agencies. Use linkages are also important. Since most rural areas of Alaska are far from cities and often have limited transportation networks, dependency on land resources for both economic and cultural purposes is high. As a result, many rural communities combine both subsistence and cash economies to survive. Three use linkage subcategories were identified by Endter-Wada and Blahna (2011): (1) open access; (2) permitted uses; and (3) illegal uses. Here, we examine permitted uses based on USFS permit data from the Tongass and Chugach NFs. We use permit data because they represent an existing secondary data set and provide information at the community level of analysis.

USFS Permit Data Systems

Community-resource linkages on public lands are not well studied, largely due to the lack of community-level socioeconomic data. Limited information presents a barrier to management designed to protect both community needs and natural resources. Monitoring resource use and the impact of use policies is especially important at the community level, where rural residents have unique cultural and natural attachments to nearby lands (Brehm et al. 2006, Kruger 2005).

The USFS has long required permits for private uses of public lands in order to facilitate, regulate, and monitor use and demand. Permit systems include written agreements that allow certain activities by specified persons under pre-established conditions and time periods. Permits also play an important role in educating permittees about the resources they access and conditions of use (Endter-Wada and Blahna 2011, USFS 2004).

Permits are required for land occupation, as well as resource uses where significant impacts are possible or where rationing of use is required. For example, each year the USFS receives thousands of requests from businesses, NGOs, municipalities, individuals, and various community groups to access NFS lands for a wide and growing range of uses. Commercial uses range from establishing infrastructure (e.g., utility corridors and telecommunications towers) and cutting timber, to commercial outfitting and guiding operations and video productions. Noncommercial uses include research and community events such as fairs and boat races (USFS 2004).

The USFS began to centralize its permit systems beginning in the late 1980s. Today, information is stored in two main systems: (1) SUDS, the Special Use Database System, which catalogs non-timber, largely non-extractive forest activities; and (2) TIM, the Timber Information Manager, which catalogs permits regarding the sale and harvest of timber and firewood. Both SUDS and TIM are linked through the USFS's corporate Infrastructure Database System – INFRA.

SUDS was developed in the 1990s to control and monitor: (1) public land that is occupied for an extended period of time; (2) commercial uses of land; and (3) group and other non-commercial uses. SUDS has two major permit categories: land use permits and recreational use permits. Both permit types can be either commercial or non-commercial. Examples of commercial land uses include utility corridors and construction sites for the lodges and cabins. SUDS permits are also issued for non-commercial land uses, including both land and recreational uses as long as the use of one permit holder does not interfere with that of another. Examples include group uses and many forms of recreation such as hiking and camping (USFS 2004). Commercial recreational use permits are often allocated to outfitter and guiding operations.

For each permit, SUDS includes information such as permittee name and residence, resource and location of use, time and intensity of use, fees charged, inspections, and permit status. SUDS data are gathered in the field or Ranger District office, and are automatically uploaded to INFRA. Beginning in 2002, the USFS expanded efforts to ensure proper data collection and recording, in part motivated by the realize action of the potential usefulness of SUDS data in resource management, agency reporting, and addressing community-level use and needs.

TIM is used store and process sale and harvest data for timber and non-timber special forest products such as burls, mushrooms, bark, berries, and many more. It was developed beginning in 1995, with full-scale implementation in 2002. TIM includes three types of permits: Free Use, Personal Use, and Commercial Use. Free Use Permits record the amount and location of firewood, saw logs, and special forest products for household consumption. Alaskan residents can harvest up to 10,000 board feet of green standing saw logs of timber/person/year without charge, and up to 25 cords of firewood/person/year (USFS 2004, Miller 2008). There are no established limits for special forest products. Personal Use Permits are also available without charge, and allow customary use and trade in traditional cottage industries for both timber and non-timber products (USFS 2007). Commercial Use Permits are issued on a fee basis to commercial harvesters that harvest timber and non-timber products for subsequent sale to third parties. Like SUDS, TIM data are entered at the Ranger District level at the time of permit issuance, providing real-time data and report-writing capabilities for monitoring revenues and harvests levels and assisting resource allocation decisions for budgeting and personnel.

Permit Data as a Tool to Describe Community Resource Access and Use

A growing number of studies have used USFS permit data to examine community-resource linkages and their implications for forest resource management and community sustainability. Sullivan (1997) used U.S. Census data and USFS permit information from Utah's Dixie NF to: (1) evaluate the use of community as an appropriate unit of social analysis; and (2) evaluate the utility of secondary data sources – mainly USFS permit data – in describing relationships between individual communities and a variety of commodity and amenity-based resources on nearby public lands in order to better integrate social science into ecosystem management. Sullivan was able to identify that local use of Dixie resources varied significantly across communities for grazing, firewood, and timber products – differences that were not discernible at the county level. The work highlighted the need for community-level data, and identified USFS permit data as a low cost, readily available information source describing local community access and use of public resources.

Oschell and Nickerson (2008) used SUDS outfitter and guide data to determine commercial recreational supply and demand within USFS Region One – i.e., NFs and Grasslands in Montana, North Dakota, and parts of Washington State and South Dakota. By comparing permit-allocated use days with actual use, they attempted to assess the recreational supply and demand but were hindered by incomplete data within SUDS. Importantly, they found that some outfitter and guide activities, such as hunting and fishing, were limited by the USFS due to insufficient information regarding resource supply and demand pressures. Their experience with SUDS found that inconsistent data recording and maintenance limited its usefulness for land management.

Charnley et al. 2008 used TIM, SUDS, and data collected by the Bureau of Land Management to examine how declining timber harvest levels under the Northwest Forest Plan affected communities using a multi-scale approach that examined community, county, state, and regional impacts. They found that the effects of declining harvests on local communities varied as a function of: (1) the importance of the timber sector in the community in the late 1980s; (2) the extent to which the timber sector depended upon local residents; and (3) the degree to which local residents depended on USFS jobs. Other factors included the level of employment diversification within a community, and distance from major urban areas.

Finally, Lilieholm *et al.* (2010) used BLM permit data on Utah's Grand Staircase-Escalante National Monument to describe how 13 small, resources dependent communities use the Monument for, grazing, fuel wood and post cutting, and special recreation uses. They found community use differences for all permits. For example, all grazing permittees were from four communities (Boulder, Esclante, Cannonville, and Kanab), but the extent of the linkages depended on the unit of measure (allotment or animal unit months) and whether or not permit numbers were standardized for community population size (to show relative use levels). This was also the first study to map the community linkages so the results could be used in planning documents.

Collectively, these studies highlight the risks of relying on county-level or higher data when exploring resource use at the community level. Indeed, while social scientists recognize community as an important unit and scale at which to study resource-dependent communities, the lack of community-level data is often limiting (Charnley et al. 2008, Beckley 1998, Blahna et al. 2003, Endter-Wada and Blahna 2011, Sullivan 1997, Lilieholm et al. 2010). Moreover, gathering community-level primary data is often prohibitively expensive, highlighting the importance of community-level permit data when studying resource linkages on public lands. The challenge is particularly acute in Alaska, where rural communities are widely scattered, remote, and operate within unique political and administrative structures.

Methods

Study Areas

The Tongass and Chugach NFs cover tens of millions of acres in Southeast and Southcentral Alaska, and are key resources for many local communities (Figure 1). In the Southeast Alaska Panhandle, federal lands comprise 91% of the area, and 77% of that is in the TNF. The TNF extends 500 miles north-to-south, and with 16.7million acres, it is by far the largest in the NFS (Schoen and Albert 2009, USFS 2007). The TNF is comprised of a narrow strip of rugged coastline and more than 200 islands bordering British Columbia, comprising six entire boroughs and three Census areas (Figure 1). It contains one of the largest intact coastal temperate rainforests in the world – an ecosystem thought to be even rarer and more threatened than tropical rainforests – and provides habitat for grizzly and black bear, mountain goat, Sitka black-tailed deer, and wolves (Alaback and Juday 1989). Their role in protecting water quality is critical to five species of wild salmon.

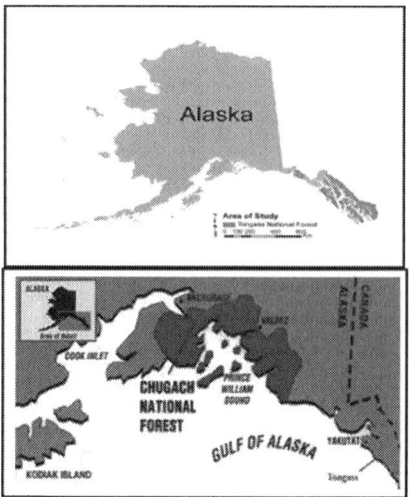

Figure 1. The Tongass and Chugach NFs.

Surrounding the TNF are 22 communities of various size and with varying levels of dependence upon the Forest (Table 1). These ranges from Juneau, Alaska's capital and regional commercial center, to small isolated Native Alaskan communities such as Angoon, Craig, Hoonah, Kake, Klawock, and Yakutat Three communities are connected to the mainland via road: Haines and Skagway to the north and Hyder to the south (USFS 2007).

	Tongass area communities				Chugach area communities				
Community	Population	Alaska Natives (%)	Median income	Popn below poverty %	Community	Population	Alaska Natives %	Median income	Popn below poverty %
Angoon	573	80.9	29,861	27.9	Anchorage-Girdwood	260,283	10.0	55,546	7.3
Coffman Cove	208	6.0	43,750	4.9	Cooper Landing	369	4.9	34,844	2.2
Craig	1,424	3.9	45,298	9.8	Copper Center	362	50.6	32,188	18.8
Elfin Cove	37	0.0	33,750	5.6	Cordova	2,454	15	50,114	7.5
Gustavus	426	8.2	34,766	14.0	Eagle River-Chugiak	29,896	0	0	55.7
Haines	1,794	18.5	39,926	7.9	Gakona	215	17.7	33,750	0.0
Hoonah	892	69.4	39,028	16.6	Homer	3,946	6.2	42,821	9.3
Hyder	98	4.1	62,034	46.7	Hope	137	5.8	21,786	11.7
Juneau	30,711	16.6	62,034	6.0	Kasilof	471	6.2	43,929	35.8
Kake	715	74.6	39,643	14.6	Moose Pass	184	0.0	87,291	0
Ketchikan	7,922	22.7	45,802	7.6	Palmer	4,533	12.5	45,571	12.7
Klawock	846	58.1	35,000	14.2	Port Alsworth	104	22.1	58,750	6.0
Meyers Chuck	21	9.5	64,375	0	Seward	2,830	20.9	44,306	10.6
Pelican	253	25.8	48,750	4.7	Soldotna	3,759	6.9	48,420	6.6
Petersburg	3,258	12	49,028	5.0	Sterling	4,702	4.4	47,700	23.8
Point Baker	8,835	8.9	51,901	7.8	Valdez	4036	10.2	66,532	6.1
Sitka	870	24.7	49,375	3.7	Wasilla	5,469	9.1	48,226	9.6
Skagway	3,722	5.1	48,420	6.6	-	-	-	-	-
Tenakee Springs	85	4.8	33,125	11.8	-	-	-	-	-
Thorne Bay	576	4.8	45,625	7.8	-	-	-	-	-
Wrangell	2,305	23.8	43,250	9.0	-	-	-	-	-
Yakutat	683	46.8	46,786	13.5	-	-	-	-	-

Table 1. TNF and CNF communities.
Source: 2000 U.S. Census

The TNF supports local communities' economies by providing a range of natural resource and activities such as fisheries, timber harvesting, recreation and tourism opportunities, and mining sites (Sisk 2009, USFS 2007). Access to fish and wildlife in particular is important in sustaining subsistence and Native lifestyles. The historic focus on timber harvests, which has greatly diminished in recent decades, was long seen as a means of providing local jobs and economic development. Indeed, the year-round nature and employment benefits of the timber sector have been important given the region's high unemployment.

Timber harvests decreased in the 1990s as a result of high harvest costs coupled with declines in global market demand, cancellation of timber sale, and national economic downturn affected the housing market (Alexander 2011). At the same time, there has been growing demand for non-extractive uses (Brooks and Haynes 1997). Since 1990, timber harvest on the TNF fell from 472, 966.25 MBF/year to 12, 667.66 MBF/year in 2012 (Alexander 2011, Warren 2010). As a result, large areas of the Tongass have been removed from the commercial timber base (Kruger and Mazza, 2006). Currently, about 57% of TNF forestland is considered productive. Of this, approximately 0.5 million acres (9%) has been harvested. The Tongass includes two National Monuments – Admiralty Island and Misty Fjords – which were designated in 1978. In 1990, Congress designated 6.4 million acres on the TNF as Wilderness – 40% of the TNF's total acreage area.

The Chugach NF is located in Southcentral Alaska, home to most of the State's population. The CNF covers 5.4 million acres and is roughly the size of New Hampshire, making it the second largest NF in the NFS (Figure 1). The Chugach is located in the mountains surrounding Prince William Sound and includes the eastern Kenai Peninsula and the Copper River Delta (USFS 2007). This region includes two national parks (NPs) (i.e., Kenai Fjords and the Wrangell-St. Elias NPs), and limited agricultural lands, oil and gas fields, and glaciers. The region's abundant lakes and streams are important spawning grounds for salmon and other species, making it popular for fishing, hunting, hiking and camping. The Forest is classified as a temperate rainforest in the Pacific Temperate Rainforest region (Bailey 2002). Compared to the TNF, CNF's forests are less suited for timber growth and harvesting.

Nineteen communities surround the CNF (Table 1). Until recently, it was generally believed that Chugach communities were more diverse in terms of socio-economic condition as compared to communities surrounding the Tongass. However, that changed with the decline of timber harvesting in southeast Alaska. The majority of Chugach communities have larger populations as compared to the Tongass. Larger and more economically diverse communities include Anchorage-Girdwood, Wasilla, Sterling, Palmer, Valdez, Homer, Soldotna, and Seward (Table 1). Anchorage-Girdwood is Alaska's largest city, with a population of 260,283—40% of the State's total. Anchorage-Girdwood is also the centre of the State's economic activities, and is situated in close proximity to the CNF. The Kenai Peninsula Borough and the Valdez-Cordova Census Area together contain 10% of the State's population (Crone et al. 2002). Small communities such as Cooper Landing, Cooper Center, Gakona, Hope, Kasilof, Moose Pass, and Port Alsowrth have populations less than 500 residents (Table 1).

Communities with high percent of Alaska natives include: Copper Center with half of its population is Alaska Natives, Port Alsworth with 22 %, Seward with 20 %, and Gakona 17 % of its population comprise Alaska Natives indicating dependence on the Chugach national forest for subsistence use and other traditional activities.

Communities with the highest annual median incomes include Moose Pass ($87,291), Valdez ($66,532), Anchorage-Girdwood ($55,546) and Cordova ($50,114) (Table 8). Moose Pass has the highest annual median income, but a population of just 206 residents and only 10.7% classified as Alaska Natives. Moreover, the town's unemployment rate is very high at 31.2% (Table 8). This disparity may reflect a large percentage of high-income retirees, although employment within the traditionally high-paying natural resources sector is also high. Finally, Table 8 shows that communities with high annual median incomes have high employment percentages in either the natural resource or service industries—a good indication of the economic importance of the forest to the local economy.

CNF management focuses on the maintenance of biodiversity and recreational opportunities, and enhancement of fish and wildlife habitat. The Forest is world-renowned for its abundant salmon populations, and the Russian River draws tens of thousands of anglers each year (USFS 2006). Sport hunters are also drawn to the Forest in search of trophy-sized moose, brown bear, Dall sheep and mountain goats, caribou, and Sitka black-tailed deer. In addition, numerous non-game species, including waterfowl and marine mammals, attract large numbers of wildlife viewers each year. CNF management works in cooperation with a host of private and public sector partners to support maintenance of hunting, fishing, and wildlife viewing opportunities (USFS 2006).

Although the Chugach is primarily a recreational forest, some timber and mining operations prevail. The CNF's timber harvest was 1,017.07MBF in 1990, but commercial harvesting was essentially eliminated in 2002 when the Chugach Forest Plan Revision Record of Decision determined that no lands were suitable for timber production (Warren 2010, ROD 2002). Mining operations include almost 3,000 small gold mining claims and a dozen gravel/stone permits (USFS 2006,). Given the CNF's limited timber and mining levels, more than 90% of the income generated by the Forest comes from campground fees, recreation fees, and mineral lease permits (USFS 2007).

Describing

Using USFS Permit Data to Describe Community Use Linkages

We used Endter-Wada and Blahna's (2011) LPL framework to explore how SUDS and TIM data could be used to describe community-resource use linkages on the TNF and CNF. We met with 11 USFS staff within the NFS Region and the Forest Supervisor Ranger District offices knowledgeable with SUDS and TIM to learn more about how permit data are gathered, entered, stored, accessed, and used, as well as some of the strengths and weaknesses of each system. Based on our conversations, we used CY 2007 data because it was the first year to have relatively complete information for both Forests.

Permits issued by the TNF and CNF were sorted by Forest and expiration date to identify permits active during CY 2007. Data were then sorted by use type, issuing Ranger District, and permittee zip code. TIM data were aggregated into three major permit categories: free use, personal use, and commercial use. To simplify the SUDS analysis, the 79 original use categories issued on the TNF and the 39 use categories issued on the CNF were collapsed into six combined-use categories across the two forests: (1) Land Occupancy and Recreational Use; (2) Outfitter and Guide Use; (3) Isolated Cabin Use; (4) ANILCA-related Use; (5) Federal Land Policy and Management Act (FLPMA) Use; and (6) Research and Educational Use (Table 2).

SUDS Combined-Use Category	Description
Land Occupancy and Recreational Use	This category includes both commercial and noncommercial land occupancies. Examples include: commercial recreational activities, filming and still photography, commercial mobile radio service, helicopter landing sites, utility corridors, parking lots, etc. Examples of noncommercial uses include all group uses (often recreation-related), community residences, and temporary land improvements.
Outfitter and Guide Use	These permits include all commercial outfitting operations that provide personal services, equipment, and materials for guests. Permittees can be both local and non-local businesses.
Isolated Cabin Use	This category includes permits for isolated recreation cabins located on sites not planned or designated for recreational cabin purposes. Most cabins originated from historic claims and, in most circumstances, these cabins are to be phased-out after 15 years.
ANILCA-Related Use	These permits are issued to rural subsistence users—mainly anglers and hunters. ANILCA use permits are unique to the State of Alaska and include: ANILCA set-net fishing camps (a commercial use but income generated is only for household consumption), and temporary hunting and fishing camp and shelter permits. All are temporary structures.
FLPMA Use	FLPMA uses include road and trail easements, grazing allotments, mining right-of-ways, sewage pipelines, etc. Land occupancy is achieved through easement or permit.
Research and Educational Use	Examples include experimental and demonstration activities, weather stations, education centers, research study sites, site survey and testing, etc.

Table 2. Combined-Use SUDS Categories Created from Original SUDS Data.

The number of permits issued to residents within a particular community was analyzed and then standardized to adjust for community size by calculating the number of permits issued to local residents on a per-1,000 household basis based on the place-level 2000 Census data:

Permits/1,000 households = (total number of permits/total number of households)*1,000

While the total number of permits is a useful measure of aggregate use, expressing use on a per 1,000-household basis gives additional insight to community dependence on public land resources. Finally, ArcGIS v9.3 was used to visually display use types and number of permits for both SUDS and TIM data at the community level.

Results

Permit Use on the Tongass and Chugach National Forests

During CY 2007, the TNF and CNF had a combined total of 3,648 SUDS permits on record, of which 2,769 were issued by the Tongass and 879 by the Chugach (Table 3).On both Forests, the largest number of permits were issued for Land Ooccupancy and RrecreationalUse categories (49.6% and 41.2%, respectively), followed by Ooutfitter and Gaides (22.5% and 31.6%), and Isolated Cabins permits (14.2% and 17.2%) (Table 3).

	Tongass		Chugach	
SUDS Permits	Active permits in CY 2007	Percent	Active permits in CY 2007	Percent
Land Occupancy & Recreational Use	1,374	49.6	362	41.2
Outfitter & Guides	623	22.5	278	31.6
Isolated Cabins	393	14.2	151	17.2
ANILCA-Related Use	208	7.5	49	5.6
Research & Educational Use	100	3.6	22	2.5
FLPMA-Related Use	71	2.6	17	1.9
Total	2,769	100	879	100
TIM Permits	Active permits in CY 2007	Percent	Active permits in CY 2007	Percent
Free Use Firewood	20	35.7	47	90.4
Free Use Sawlogs	20	35.7	3	5.8
Personal Use Firewood	12	21.4	1	1.9
Commercial Firewood	4	7.2	1	1.9
Total	56	100	52	100

Table 3. Active permits on the TNF and CNF in CY 2007.

There were far fewer TIM permits than SUDS permits in CY 2007 (Table 3). Free Use Firewood permits were well represented on both Forests. On the TNF, Free Use Sawlogs and Personal Use Firewood were also important. Reflecting its legacy of commercial timber harvest, 7.2% of the TNF's TIM permits were commercial firewood permits compared to only 1.9% for the CNF (Table 3). (It is important to note, however, that these permits set a ceiling on use – e.g., 25 cords/person/year – and thus are unlikely to capture the actual harvest level.) Overall, Table 3 demonstrates the higher demand for SUDS-related uses on both Forests – this finding is highlighted when one considers that TIM data represent a one-time (i.e., annual) use, while many SUDS permits are issued for ongoing activities.

Zip code information indicates that 79% of permits were issued to permittees residing in Alaska (78% and 82%, respectively, for the TNF and CNF). The remaining 21% of permittees were distributed across 40 states, including New Jersey, Washington, and California (5%, 4% and 3% of total, respectively). Two percent (46 permits) were issued to permittees residing in Canada. All TIM permits were issued to permittees residing within study area communities, indicating more localized use – an expected result as these permits specify the physical removal of forest products, an activity suited to on-site processing by local residents. Many of the permits issued to non-local permittees were for outfitter and guide and recreation residence permits.

Community Distribution of TNF Permittees

Overall, the study area's larger communities received the most SUDs and TIM permits – Juneau (547 permits), Ketchikan (354 permits), Petersburg (221), Sitka (188), Wrangell (181), and Yakutat (134) (Figure 2a). Compared to other studies of resource use permits (e.g., Lilieholm et al. 2010), these numbers indicatee strong use linkages with the TNF. Communities with the most TIM permits included Petersburg (18 permits), Ketchikan (12), Juneau (6), Tenakee Springs (5), and Wrangell (5) (Figure 2a). Not surprisingly, Petersburg and Ketchikan represent the region's historic wood processing centres.

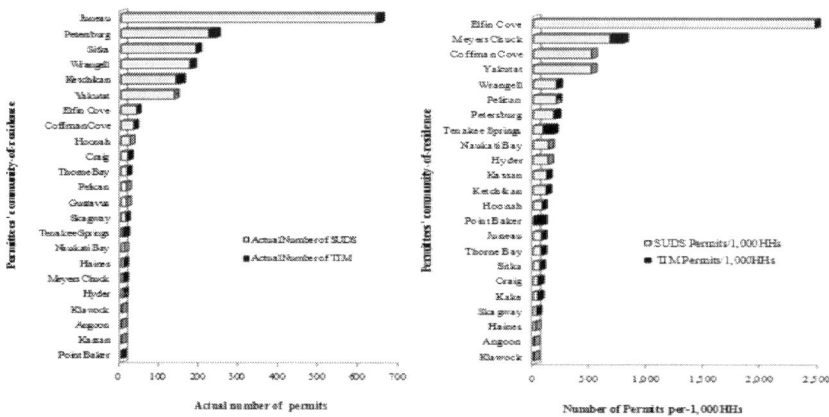

Figure 2. SUDS and TIM permits active in CY2007 by TNF community expressed in: (a) totals; and (b) on a per-1,000 household basis.

Figure 2b presents SUDS and TIM data adjusted on a per-1,000-household basis to portray a dependency-based view of resource-use linkages. Here, the communities of Elfin Cove, Meyers Chuck, Yakutat, and Coffman Cove ranked high on permit use on a per-1,000-household basis (Figure 2b). Similarly, there were large numbers of TIM permits in Meyers Chuck (111 per 1,000 households), Tenakee Springs (85), Point Baker (77), and Edna Bay (53) (Figure 2b). In general, smaller communities had more SUDS and TIM permit on a per-1,000-household basis than larger communities (see Table 1).

SUDS Permits for the TNF – Figures 3a through 3c show permit use on a per-1,000-household basis for three major types of SUDS permits – Outfitters and Guides, ANILCA-related uses, and Land Occupancy and Recreational Use. For Outfitter and Guide permits – one of the most commonly issued permits on the TNF – important communities include Elfin Cove, Yakutat, Tenakee Springs, and Wrangell (Figure 3a). Many other communities also had some level of Outfitter and Guiding activity based on CY 2007 data (e.g., Sitka, Hyder, and Petersburg).

Another important permitted use on the Tongass is ANILCA-related use. These permits track subsistence activities such as fishing and hunting, and are of special interest to Tongass managers given existing laws and mandates to ensure the continuity of subsistence lifestyles. As shown in Figure 3b, Yakutat and Pelican were relatively high-use ANILCA-related communities, and both communities have high percentages of Alaska Natives – 47% and 26%, respectively.

Land Occupancy and Recreational Use is another important use on the TNF. Here, high-use communities include Elfin Cove, Meyers Chuck, and Coffman Cove (Figure 3c). Intermediate-use communities include Yakutat, Hyder, Kassan, and Petersburg. Across all three permits types, Elfin Cove is more of a 'generalist,' having a diverse array of resource use linkages with the TNF (Figures 3a through 3c). In contrast, Meyers Chuck is linked to the TNF in a more specific way – e.g., all SUDS permits issued to residents in Meyers Chuck were for Land Occupancy and Recreational Use.

(a) 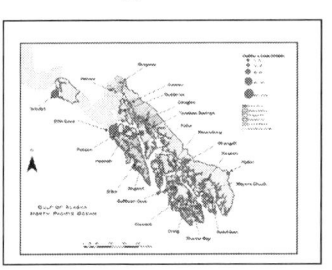 (b)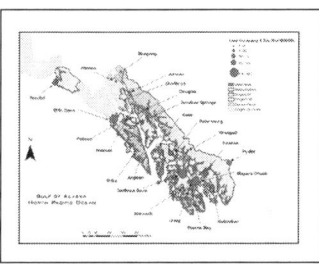

(c)

Figure 3. SUDS permits for Tongass-area communities on a per-1,000-household basis for: (a) Outfitters and Guides; (b) Land Occupancy and Recreational Use; and (c) ANILCA use permits.

Specific TIM Permits for the TNF –The majority of TIM permits issued on the TNF were for Free Use Firewood and Free Use Sawlogs (Table 3). On a per-1,000-household basis, the dominant users of TIM permits summed across all categories were Meyers Chuck, Tenakee Springs, Point Baker, and Edna Bay (Figure 4). Like many communities in the study area, they are remotely located and do not have road access to larger cities, which may be one reason why they appear to depend heavily upon firewood for heating.

Ketchikan was the only community that was issued Personal Use Firewood permits, and Wrangell was the only community where Commercial Use permittees resided (data not shown). These two communities have a history of extractive forest use and were home to southeast Alaska's timber sector, where mill operations began in 1956 (Alexander 2011, Alaska Forest Association 2009, Halbrook et al. 2009). Although Ketchikan's pulp mill closed in 1997, the community continued to support several sawmills even after the TNF's shift in management during the 1990s. Currently, a few small sawmills remain in Ketchikan, Wrangell, and Klawock, including a veneer plant in Ketchikan (Kilborn et al 2004, Alaska Department of Commerce 2009).

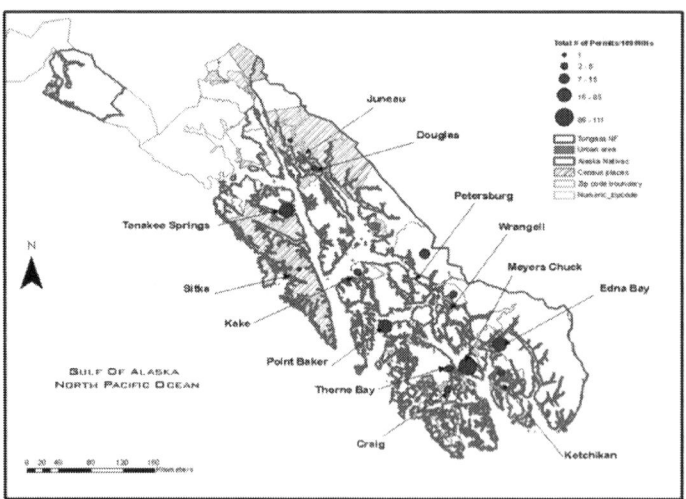

Figure 4. Tongass TIM permittees' community-of-residence.

Community Distribution of CNFPermittees

For the Chugach study area, the largest number of SUDS permits were issued to residents in Anchorage-Girdwood, Cordova, Wasilla, Cooper Landing, and Homer, respectively (Figure 5a), indicating strong use linkages between these communities and the CNF. Most TIM permits were issued to residents of Seward, Anchorage-Girdwood, Moose Pass, and Copper Center.

Palmer and Sterling also had sizable numbers of TIM permittees, while a few were issued to residents in Eagle River-Chugiak (Figure 5a). As described above, all TIM permits were issued to study area communities – once again reinforcing the local nature of forest products use.

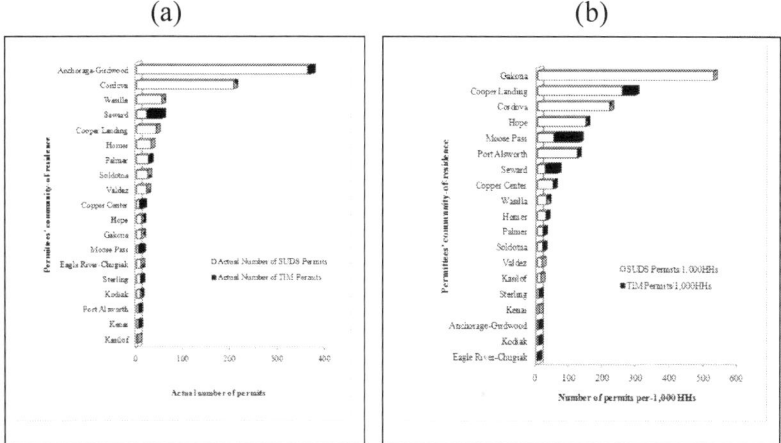

Figure 5. SUDS and TIM permits active in CY 2007 by CNF community expressed in: (a) totals; and (b) on a per-1,000 household basis.

Figure 5b shows SUDS and TIM permits on a per-1,000-household basis. Communities with the highest number of SUDS permits included Gakona, Cooper Landing, Cordova, Hope, and Port Alsworth, respectively (Figure 5b). For TIM permits adjusted on a per-1,000-household basis, Moose Pass, Cooper Landing, and Seward were prominent – small, isolated towns with heavy dependence on firewood for home heating use (U.S. Census). As with the TNF region, expressing permit use on a per-household basis presents a very different view of community resource use linkages. In short, while residents of major population centers receive the largest numbers of permits, when adjusted for population size, small communities access resources at much higher per person use rates – indicating a higher level of dependence on resource use permits.

SUDS Permits on the CNF – Large variations in SUDS use patterns between communities were found in the CNF region. For example, on a per-1,000-household basis, Cooper Landing and Hope had the largest number of Outfitter and Guide permits (Figure 6a), followed by Gakona and Port Alsworth. For Land Occupancy and Recreational Use permits, Cooper Landing dominated, followed by Cordova, Girdwood and Seward (Figure 6b). Although there were not many ANILCA use permittees on the CNF, the data indicate that two communities, Girdwood and Cordova, the former located within the Anchorage-Girdwood periphery, are areas of residence for the majority of ANILCA permittees (Figure 6c)

As with the TNF, the data reveal some communities to be specialists, while others were more diversified with respect to SUDS permit types. For example, Anchorage-Girdwood, Cordova, and Homer had more than one permit type, indicating more diverse resource use linkages than other communities. Despite this, the proportions of use permits vary from community-to-community. For instance, in Cooper Landing there were more Outfitter and Guide permits per-1,000 households than Land Occupancy and Recreational Use permits. On the other hand, in Anchorage-Girdwood, there was almost an equal proportion of the various SUDS permits expressed on a per-1,000 household basis. In Cordova, there were more Isolated Cabin permits as compared to Land Occupancy and Recreational Use and Outfitter and Guide permits. In Anchorage-Girdwood, Homer, Seward, Soldotna, and Wasilla, there were relatively fewer of these permits expressed on a per-1,000-household basis. Cordova and Homer were the only communities that had research and educational use permits (data not shown). Overall, we found that communities are differentially linked to the CNF, with many linked by more than one use type. Such information can assist managers in estimating the extent of community use linkages for different types of permitted activities.

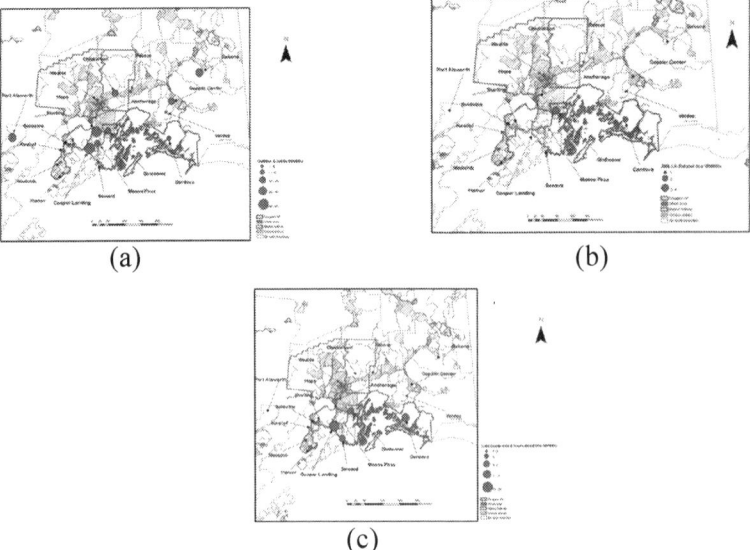

Figure 6. SUDS permits for Chugach-area communities on a per-1,000-household basis for: (a) Outfitters and Guides; (b) ANILCA-related use; and (c) Land Occupancy and Recreational Use permits.

TIM Permits on the CNF – There were far fewer TIM permittees on the CNF as compared to the TNF in 2007. Moreover, 95% of TIM permittees in the CNF were for Free Use Firewood permits. The majority of TIM permittees resided in Moose Pass, Cooper Landing, Seward, and Girdwood (Figure 7). The small number of TIM permits on the CNF corroborates that the CNF is managed primarily for recreation rather than timber harvesting.

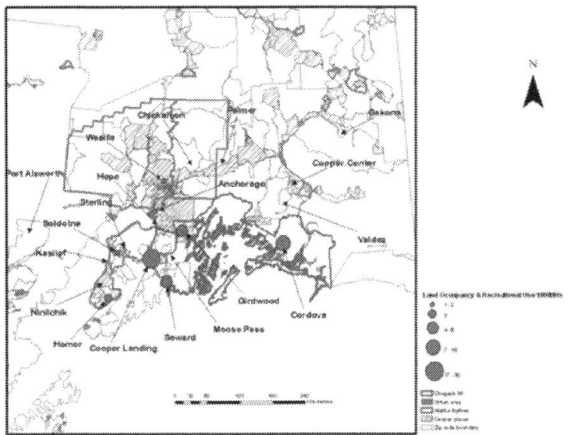

Figure 7. Chugach TIM permittees' community-of-residence.

Discussion

By utilizing permit data in this analysis, it was possible to gain insight into local communities' use of forest land and resources by way of permittee addresses. The addresses were determined by the zip code entered for each permit. This allowed us to trace permitted use activities on the forests back to the community level of analysis, and the number and type permits were then tallied and mapped for each community. This method provides a more direct measure of community-forest linkages than standard social assessment measures, and addresses the regional scale of analysis based on the location of communities being mapped.

Based on 2007 permit data, land occupancy for both commercial and noncommercial use activities is by far the largest permit-based resource use on the TNF (Table 3). In contrast, there were fewer permittees directly extracting forest products, and these were primarily firewood and saw logs (Table 3). The majority of SUDS permittees resided in Juneau, Ketchikan, Petersburg, Sitka, Wrangell, and Yakutat (Figure 2), although the number of permits per 1,000 households was found to be higher in communities with smaller populations. Expressing permit use on a per-household basis presents a very different view of community-resource use linkages.

Most TNF TIM permittees resided in Juneau and Ketchikan (Figure 2a). Again, these are larger communities where the actual numbers of permits are large, but may not indicate the relative level of linkages to the forest (Table 1). On a per-1,000 household basis, Meyers Chuck, Tenakee Springs, Point Baker, and Edna Bay emerged as resource-dependent communities (Figure 2b). Common characteristics of these communities include: (1) relatively small population sizes; (2) large percentages of Alaska Natives; (3) remote locations; (4) farther from major airport/ferry/float plane service; and (5) compared to Juneau and Ketchikan, they have higher employment in service and natural resource-related sectors (Table 1). All these socioeconomic attributes positively correlate with strong resource-use linkages.

We found that the TNF issued relatively low numbers of permits for wood harvesting, and more permits were issued to harvest firewood than timber. These findings are consistent with changes in TNF management from a timber-oriented forest to an amenity-based forest. Indeed, a shift in forest management policy under the 1997 TNF management plan made four main regulatory changes regarding timber harvests: (1) a change enacted in 1999 to remove 100,000 acres from the harvestable timber base, reducing it from 676,000 to 576,000 acres; (2) the allowable harvest rotation age was doubled to 200 years, making it harder to develop a forest products sector based on second-growth timber; (3) allowable road density in the Forest was reduced from 1 mile/mile2, to 0.7 mile/mile2 of forested land, compounding access issues already found on the Forest; and (4) the average allowable sale quantity or ASQ was reduced from 267 MMBF to 187 MMBF, potentially limiting the annual production of the forest.

After adoption of the 1997 TNF plan, timber sales became a by-product of ecosystem management (USFS 2006). This, combined with global competition and high production costs, lead to a large decline in timber processing capacity over the last 15 years (Clausen and Schroeder 2004, Crone 2004, Tsournos and Haynes 2004, Braden et al. 2000). Today, there are just a handful of small sawmills operating in TNF communities – namely Ketchikan, Wrangell and Klawock (Alaska Department of Commerce 2009). Indeed, there were no commercial sawlog permits active in 2007.

On the CNF, there were more permits for land occupation and recreational uses than timber harvesting or other extractive uses. Commercial outdoor recreation – particularly outfitter and guide permits – were by far the largest recreational activities. Other recreational and noncommercial activities included cabin and campground permits. These findings for the CNF are consistent with the Forest's emphasis on recreation over timber and other forest products.

The majority of SUDS permittees resided in large CNF communities like Anchorage-Girdwood and Wasilla, a suburb of Anchorage (Figure 5). These communities have relatively large populations and diverse economies. As discussed above, having large numbers of permits alone does not necessarily indicate a strong linkage to the Forest. Indeed, on a per-1,000 household basis, communities such as Gakona, Cooper Landing, Cordova, Hope, Port Alsworth, and Girdwood were identified as communities having strong per-1,000-household linkages to the CNF (Figure 5).

There were many TIM permittees from Seward, Anchorage-Girdwood, Moose Pass, and Copper Center, respectively (Figure 5). Overall, Land Occupancy and Recreational Use and Outfitter and Guide permits are the largest use categories on the CNF, a similar finding with that of the TNF (Figures 3 and 6).Communities highly dependent on direct forest products include: Moose Pass, Seward, Cooper Landing, and Gridwoood (Figure 5). The majority of TIM permittees held permits for free use firewood, an indication of a strong direct resource-use linkage to the CNF. However, there were fewer TIM permittees in the CNF compared to the TNF (Table 3). Overall, on both forests, free use firewood permits represent more than 90% of the TIM permits – a strong indicator of local communities' linkages to both Forests.

In our two study areas, rural communities met their needs through a variety of strategies. Some are generalists engaging in a wide range of activities, while others may specialize in niche areas where existing assets provide an advantage. Such differences in livelihood strategies, and implications for social assessment for NFS management, were evident through the community analysis of the numbers and types of permits issued on the TNF and CNF. Understanding these differences are also critical for identifying the social impacts of agency management and policy changes, such as reduced harvests, limiting recreation access or non-timber forest extraction, closing or changing road access, and many more. The essence of social assessment and impact analysis is understanding the distributional effects of changes in forest policy or management.

While we may not be able to confidently describe between Forest comparisons yet (due to inconsistencies with data collection), the relative comparisons within forests show the potential for using this as a community-resource use linkage social assessment tool for individual forests, and the potential for cross Forest and agency-wide values of resource use access if the agency implements more consistent data collection methods. Moreover, the results provide a map display for easy visual representation and use in planning.

Permit Data Limitations

Scale is an important factor in defining community-resource use linkages and dependency (Beckley 1998, Blahna et al. 2003, Morse et al. 2009). Human communities are defined more by social interaction than by any political or administrative boundaries. Thus it is difficult to clearly identify social community boundaries and relate data collected at different scales of analysis. We used zip code combinations that best reflected the boundaries of cities and towns because they represent a unit of analysis that is easily understood and is used in the USFS's planning process. Data collected at the zip code level, for example, may encompass multiple towns in rural areas, or high population areas may contain several zip codes. For Alaska the problem is reduced somewhat because most communities are rural and tend to be geographically dispersed. When zip codes were not contained fully within the boundary limits of one municipality, we included zip codes if more than 50% of the zip code area was found within the community place. This was rare, however, and was usually an issue in high population areas of Juneau and Anchorage.

Perhaps the biggest weakness with the application of this method is incomplete or inconsistent collection of permit data. While INFRA was operational in 1997, the first year that relatively complete data were available for both Forests was 2007. Therefore, we only have a snapshot in time, and we cannot compare trends which may be the most valuable use of these data for the assessment of social linkages. And even by 2007 we found problems in the collection and storage of permit data, especially for SUDS. For example, information on permittee place-of-origin was sometimes not recorded correctly. In other cases, some SUDS permit holders may have resided outside of Alaska for part of the year, but they have local business addresses. In those situations, this analysis might have overestimated community-use linkages because those permittees may not truly represent local communities as typically defined.

Many data-entry problems we found originated at the Ranger District level and probably reflect lack of adequate staff or training. Many Ranger Districts have experienced reduced staffs and budgets in recent years. There are also differences in reporting procedures due to the lack of standardized systems. Even though every region receives the same directive from the USFS Washington Office, Region may implement the guidance slightly differently, and then individual NFs may implement Regional guidance differently. A study by Oschell and Nickerson (2008) reported similar problems in their needs assessment study of outfitter and guide permits for USFS Region One. Such variations make comparisons across Forest and Regions difficult, and even the results for individual Forests must be considered estimates of permitted activities.

Finally, based on discussions with USFS employees, we learned that some residents access forest resources without permits. Unfortunately, both the TNF and CNF lack sufficient resources to monitor and control many of the illegal forest uses. There are also many extractive activities that do not require a permit, such as collecting mushrooms, burls, barks, cones, and other products. For some activities, permits are optional or collected inconsistently, such as Christmas tree cutting, camping, and nature center and trailhead access. Thus, permit data only provide estimates of community-forest use linkages described by Endter-Wada and Blahna (2011). But currently they are the best data available, and hopefully, by illustrating the potential value of these data, it will lead to improved permit recording and analysis procedures.

Conclusion

SUDS and TIM permits are currently used by the USFS to summarize forest product collections and activities on each NF. This analysis shows that permits can also be used as social science data to help identify community-resource use linkages that may be particularly sensitive to forest management and policy changes. Since permit data are already collected on NFs and stored in an agency-wide centralized database and mapping community linkages is relatively easy. This would be a practical and low-cost approach to provide social science data that are relevant for forest planning. More importantly, however, permit data can be used as an input to illustrate the linkages between forest management actions and community resilience. Additionally, these data provide more direct forest-community linkage information than standard social assessment measures that focus on community characteristics that may or may not be affected by changes in forest management practices (Endter-Wada and Blahna 2011). This is especially true in Alaska. Indeed, given the rural nature of the state, the predominance of public lands, and the rural and subsistence lifestyles of many Native and non-Native communities, understanding community-resource use linkages is paramount to effective planning and management.

Lesson Learned and Method Application in the Ethiopian Context

In Ethiopia, land is owned and administered by the government whereby its 90 million people must have to sign a long-term lease agreement to access land and/or ensure their use right. Currently, the government programs guided by such a principle of strong regulation of individual property rights and successive control over tenure relations altered the role of customary resource use rights and land administration. Historically, such continuous change and lack of clarity regarding the traditional use rights of 85% of rural residents which include peasants, pastoralists and others has impacted their historical and customary right to access and use land, forest, water, and mineral resources upon which they depend on for their livelihoods.

Therefore, the method used in Alaska can be useful for the current situation in Ethiopia and it is possible to adopt the approach to understand and assess how the rural people access to the land and may have impacted by the current land management policy. Furthermore, the advantage of the method is multifaceted because collecting primary data for a similar research is very much time consuming and expensive for a country like Ethiopia. Therefore, using the available secondary data sources (e.g., records of income and sales tax collected, list of profile of lease holders, and other similar government records) it is possible to capture how people access and use land and potential policy impacts on the rural residents. Currently, the Ethiopian government has built a centralized database on income and sales tax records which reflects various socioeconomic activities, justify the need of this kind of study as timely.

References

Alaback, P.B., and G.P. Juday. 1989. Structure and composition of low elevation old growth forests in research natural areas of southeast Alaska. *Natural Areas J.* 9:27-39.

Alaska Department of Commerce, Community and Development. 2009. *Information on southeast Alaska saw mills.* http://www.commerce.State.ak.us/10, [Accessed December 15, 2009].

Alaska Division of Forestry. 2010. *Statewide Assessment of Forest Resources*. State of Alaska.80 p.

Alaska Division of Parks and Outdoor Recreation.2002. *Alaska Forest Legacy Program Assessment of Need*. Alaska Division of Parks and Outdoor Recreation. 87p.

Alaska Forest Association. 2009. http://www.akforest.org/index.html, [Accessed September 02, 2009].

Allen, S. D., G. Robertson, and J. Schaefers. 1998. *Economies in transition: An assessment of* trends relevant to management of the Tongass National Forest. USFS-PNW-GTR-417. Portland, OR: Pacific Northwest Research Station.

Allexander, J. S. 2011. The Status of Tongass National Forest 2009. USDA-Forest Service, Alaska Region. R10-MB-749.

Arnold, M. and I. Townson. 1998. *Potential of forest product activities to contribute to rural incomes in Africa.* Natural Resource Perspectives 37. London, UK: Overseas Development Institute.

Bailey, R.G. 2002.Ecoregion-Based Design for Sustainability, New York, Springer-Verlag.

Beckley, T. 1998. The nestedness of forest dependence: A conceptual framework and empirical exploration. *Society Nat. Resource.* 11(2):157-168.

Berger, T. R. 1985. *Village journey: The report of the Alaska Nativew Review Commission.* New York: Hill and Wang.

Blahna, D. J., D. Carr, and P. Jakes. 2003. Using social community as a measurement unit in conservation planning and ecosystem management. In *Understanding Community Forest Relations,* tech. ed. L. Kruger.USFS-PNW-GTR-566:59-80. Portland, OR: Pacific Northwest Research Station.

Braden, R., K. Cunningham, B. Lipke, and I. Eastin. 2000. *An assessment of market opportunities for Alaskan forest product exports.* USFS- PNW-GTR-500. Portland, OR: Pacific Northwest Research Station.

Brehm, J. M., B. W. Eisenhauer, and R. S. Krannich. 2006. Community attachments as predictors of local environmental concern: The case for multiple dimensions of attachment. *Amer. Behavioral Scientist* 50(2):142-165.

Brooks, D. J., and R. W. Haynes. 1997. *Timber products output and timber harvests in Alaska: Projections for 1997-2010.* USFS- PNW-GTR-409. Portland, OR: Pacific Northwest Research Station.

Brooks, D., and R. Haynes. 2001. *Recreation and Tourism* in southcentral Alaska: Synthesis of recent trends and prospects. USFS-PNWR-GTR-511.Gen. Tech. Rep. PNW-GTR-409. Portland, OR: US Forest Service, Pacific Northwest Research Station.

Carroll, M. S. 1991. *Northwest loggers: An occupational community at risk in the forest management wars.* Presented at the annual meeting of the Rural Sociological Society, August, Columbus, OH.

Charnley, S. 2006. *Northwest forest plan: The first ten years (1994-2003): Socioeconomic monitoring results.* 6 Vols. USFS-PNW-GTR-649. Portland, OR: Pacific Northwest Research Station.

Charnley, S., E. M. Donoghue, and C. Moseley. 2008. Forest management policy and community well-being in the Pacific Northwest. *J. For.* 106(8), 440-447.

Clausen, D. L., and R. F. Schroeder. 2004. *Social acceptability of alternatives to clearcutting; Discussion and literature review with emphasis on southeast Alaska.* USFS-PNW-GTR-594.Portland, OR: Pacific Northwest Research Station.

Crone, K. L., P. Reed, and J. Schaeferes. 2002. *Social and economic assessment of the Chugach National Forest area.* USFS-PNW-GTR-561. Portland, OR: Pacific Northwest Research Station.

Deal, R. L., B. Cochran, and G. LaRocco. 2012. Bundling of ecosystem services to increase forestland value and enhance sustainable forest management. *Forest Policy and Economics,* 17, 69-76.

Endter-Wada, J., and D. J. Blahna.2011. Linkages to public land framework: toward embedding human into ecosystem analyses by using "inside-out social assessment". *Ecological Application.* 21(8), 3254-3271.

Fitchen, J. M. 1991. *Endangered spaces, enduring places: Change, identity and survival in rural America.* Boulder, CO: Westview.

Gallagher, T. J. 1988. Native participation in land management planning in Alaska. *Arctic* 41(2), 91-98.

Geisler, C. C., and H. Mitsuda. 1987. Mobile- home growth, regulation, and discrimination in Upstate New York. *Rural Soc.*52,532 543.

Gunderson, L., and C. S. Holling. 2002. *Panarchy: Understanding transportations in human and natural systems.* Washington, DC: Island Press.

Halbrook, J. M., T.A. Morgan, J. P. Brandt, C. E. Keegan III, T. Dillon, and T. M. 2009.*Alaska's timber harvest and forest products sector, 2005.* USFS-PNW-GTR-787. Portland, OR: Pacific Northwest Research Station.

Humphrey, C., G. Berardi, L. Fortman, C. Geisler, C. Johnson, J. Kusel, R. Lee, S. Macinko, M. Schulman, and P. West. 1993. Theories in the study of natural resource dependent communities and persistent rural poverty in the United States. In *Persistent poverty in rural America,* Rural Studies Series, 136-172. Boulder, CO: Westview.

Ives, E. D. 1988. *George Magoon and the Down East Game War: History, folklore and the law.* Chicago: University of Illinois Press.

Jakes, P., T. Fish, D. Carr, and D. Blahna. 1998. Functional communities: A tool for national forest planning. *J. Forestry,* 96 (3),33-36.

Kennedy, J. J., and N. E. Koch. 2004. Viewing and managing natural resources as human-ecosystem relationships. *For. Policy and Econ.* 6, 497-504.

Kilborn, K. A., Parrent, D. J., Housley, R. D. 2004. Estimating Sawmill Processing Capacity for Tongass Timber. USFS-PNW-RN-545.

Knutsson, P., and M. Ostwald. 2006. *A process-oriented sustainable livelihoods approach: A tool for increased understanding of vulnerability, adaptation and resilience.* Mitigation and Adaptation Strategies for Global Change. DOI 10.1007/S11027-006-4421-9.

Kruger, L. 2005. Community and landscape change in southeast Alaska. *Landscape and Urban Plan.* 72, 235-249.

Leach, M., R. Mearns, and I. Scoones. 1999. Environmental entitlements, dynamics and institutions in community-based natural resource management. *World Development,* 27, 255-247.

Lee, R. 1991. Moral exclusion and rural poverty: Myth management and wood products workers. Presented at the Annual Meeting of the Rural Sociological Society, August, Columbus, OH.

Lilieholm, R.J., M.E. Tessema, D.J. Blahna, and L.E. Kruger. 2010. Using Secondary Data to Estimate Community-Resource Linkages in Utah's Grand Staircase-Escalante National Monument. Pages 424-435 in Proceedings of the Grand Staircase-Escalante National Monument Science Symposium, Southern Utah University. 545 pages.

Oschell, C., and N. P. Nickerson. 2008. *Baseline information for regional needs assessments.* Missoula, MT: Institute for Tourism and Recreation Research.

Schoen, J., and D. Albert. 2009. *Southeast Alaska Conservation Assessment.* Information retrieved online on 09-11-09 from http://ak.audubon.org, accessed February 16, 2010.

Sisk, J. 2009. *An introduction to the industrial history of southeastern Alaska: A geographic perspective in southeast Alaska conservation assessment.* http://home.gci.net, [Accessed September 11, 2009].

Southeast Alaska Conservation Council. 2009. Information retrieved from http://www.seacc.org/successes/tongass-history, [Accessed August 24, 2009].

Stedman, R. C. 1999. Sense of place as an indicator of community sustainability. *Forest Chronicle,* 75(5), 765-770.

Stedman, R. C., J. R. Parkins, and T. M. Beckley. 2004. Resource dependence and community well-being in rural Canada. *Rural Soc.* 69(2), 213-234.

Sullivan, M. 1997. *Monitoring forest resource dependence in southern Utah: Applications to ecosystem management*.MS thesis, Utah State University, Logan.

U.S. Census Bureau. 2000. Alaska population and household place-level data. http://factfinder.census.gov, [Accessed May 01, 2006].

U.S. Department of Transportation. 2002. *Highway statistics 2001.* Federal Highway Administration, Office of Highway Policy Information. [online] URL: http://www.fhwa.dot.gov/ohim/hs01/aspublished/, [Accessed April 07, 2008].

USFS. 1997. Tongass Land Management Plan revision. Ketchikan, AK: United States Forest Service.

USFS. 2004. *Forest Service Special Use Handbook: Application and Authorization Process*. Ketchikan, AK: United States Forest Service.

USFS.2006. Tongass National Forest. Information retrieved from http://www.fs.fed.us/r10/tongass, [Accessed June11, 2008].

USFS. 2007. Tongass National Forest Land and Resource Management Plan: Draft Environmental Impact Statement. Ketchikan, AK: United States Forest Service.

Warren, D. D. 2010.*Production, price, employment, and trade in Northwest Forest sectors, all quarters 2009*, USFS-PNW-RB-259. Portland, OR: Pacific Northwest Research Station.

Wolf, J. W. 1998. Subsistence economies in rural Alaska: Crises on the last frontier. *Cultural Survival Quarterly,* 22(3).

CHAPTER 20:

Conceptualising Power and Empowerment in the Context of Public Participation in Urban Development Projects in Sub-Saharan African Cities.

Mentesnot Mengesha & Mammo Muchie

INTRODUCTION

The rural population is declining across the world while, in the same proportion, the urban settlement is on the increase. New cities are emerged; the existing ones are expanding in space and infrastructure to accommodate the new comers. In some parts of the world, cities are joining the megacity status currently in Asia and Latin America. These are happening as result of urban developments. When people are living in a defined spatial space, it is natural that they will react to the stimuli of their socio-economic environment. In many cases residents like to engage at different levels in issues that directly affecting them.

My aim here is to discuss how the power holders and the public positioned themselves in the decision making process while such fast evolving urban development is taking place. The power relationship and the empowerment process are at times complex that result detrimental effect on the sustainability of area-based infrastructure or community development projects in sub-Saharan African cities. The beneficiaries of small area-based or neighbourhood urban development projects need to clearly understand from where power emanates and how it is used as well as how empowerment balances power and then help in enhancing a meaningful participation in the decision making process. This is because in many sub-Saharan African countries the political elites have more decision making power. In the following section, I will discuss the discourses around power and empowerment in relation to public and community participation in the decision making process which could be applicable when urban development projects are planned and undertaken in fast growing African cities like Addis Ababa of Ethiopia.

Power in the context of participation

The power relationships between the public and other stakeholders in the urban development projects are matters that should be taken into account in any decision making process. It is essential to fully understand the dynamics between these forces including the sources and functions of power and their impacts on participation in the process of urban development projects. Initially, I will discuss different thinkers' perspectives on power which will lead to a discussion of the related concept of 'empowerment'. I will explore the two concepts and their functions in a decision making process.

Power in the context of public participation is a complex concept. It carries different and sometimes conflicting meanings when viewed from different philosophical and ideological perspectives. Some agree that power implies a level of authority 'to limit the range of actions that others can perform, or constrain their choices'; therefore, it requires a level of interaction with others to manifest itself (Lukes, 2005:73; see also Haugaard, 2002). Such a notion arises from the conception that power is emanating from 'a central, symbolic place or position in society' (Newman: 2004:139). The manifestation of power is also seen through the interaction process of the powerful, notably City Council officials, and the powerless, constructed in a way that power is to be given and received from a centralised place in a society. In this way of thinking, centralised power is considered as more symbolic, 'functioning as a way of organising power relations around sovereign institutions and laws' (Haugaard and Lentner, 2006:171). Marxist thinkers also agree on the notion of the centrality of power and often consider it as 'an oppressive and illegitimate arrangement' (Dahl, 1989). For the Marxists, the legitimate arrangement is the centrality of proletarian power. Hence the centrality of power remains the same except that the sovereign is replaced by the proletariat.

Perhaps it is Foucault who has done most to dispute the notion of power as a sovereign, unitary, centralised and repressive construct. He argued that 'power permeates at all levels of society' (Fox, 1998:416). Foucault differs in three main areas from other thinkers. Firstly, according to Foucault, power is not a negative and prohibitive phenomenon, rather it is active. Secondly, power does not originate from one particular source – it emanates from many sources. Thirdly, Foucault indicates that power is not held by any one person, group or class. Hence, power is neither a possession nor something that is acquired. Therefore, Foucault summarises power as a relationship or a network of relationships, distributed throughout society, affecting individuals in various ways. However, Foucault acknowledges that 'the effects of power ultimately support a certain social class, regime and economic system' (Holub, 1985:250).

Lukes (2005: 29), on the other hand, summarises three conceptual views that explain what power is. These are the pluralist view (which he calls the one dimensional view); the view of critics of pluralism or elitist (the two dimensional view); and a third view of power (which he calls the three dimensional view). Accordingly, in the first dimension, power is to do with 'the use of superior resources (by A: the relatively powerful) to reward or punish the behaviour of those with fewer resources (B: the relatively powerless)'. Hence power is manifested when resources are used by the relatively powerful 'to overtly coerce (B) to do what (A) wants' (Dahl 1969; Culley and Hughey, 2008: 101-102). The first dimension concerns observable decision-making behaviour in societies, especially when this involves deciding between contested policy options; that is, how are decisions taken and conflicts resolved? (Kelly, 2006: 2119). This is a zero-sum assumption, as what the powerful achieves and gains is at the expense of the powerless. In this particular view, power could emerge from personal skills, technical expertise and knowledge, money or wealth to be used over those with limited personal skills, knowledge or money.

Unlike the first dimension, the second dimension of power is generally understood as the ability to determine who participates and what is debated in decision making about key issues. This dimension manifests itself through setting agendas or constructing barriers to participation by preventing the less powerful from raising issues, resulting in their withdrawal from participation (Bachrach and Baratz 1992; Culley and Hughey, 2008: 101-102). The relationship between parties reflects one in which the 'rules of the game' (e.g. institutional procedures or agendas) are used to systematically benefit one group (A) over another so that one group (B) is less able to defend and promote its interests. Therefore, the second dimension concerns the determination of which issues or options are presented to decision makers; that is, how is it decided which issues are included on the decision-making agenda and which issues are not (Kelly, 2006: 2119; Bacharach &Baratz, 1970).

Steven Lukes (1974) discusses the first two dimensions of power i.e. pluralist and elitist views before he argues for the third dimension. Firstly, according to pluralists like Robert Dahl (1956) power is manifested as the relative influences of various interest groups over executive decision-making. Secondly, according to elitist theorists like Mills (1956) power is the ability to set the agenda prior to any decision making taking place. Although there are some similarities (but for different reasons) between the elitist and pluralists' models mentioned above, Lukes argues that power has a third dimension which is the ability to shape an agent's preferences or perception of their 'real interest'. Lukes suggests that the powerful can control the weak by influencing their 'real interests' (Ron, 2008). Lukes distinguished between 'perceived' and 'real' interests and suggested that the powerful can control the weak by causing them to misperceive their real interests (Young, 1978). Their influence includes the power to prevent the formation of grievances by shaping perceptions, cognitions, and preferences in such a way as to ensure the acceptance of a certain role in the existing order. Whereas the pluralist and elitists models assumed that agents can always identify and articulate their own interests, Lukes' view of the third dimension of power refused to take that for granted (Heyward, 2007:48).

The third dimension according to Lukes is generally thought to manifest itself as the ability of the relatively powerful (A) to control and disseminate myths and ideology which are used to shape the very thoughts, desires and interests of the relatively powerless (B) (Gaventa 1980; Lukes 1974; Parenti 1978). The perception of what is possible or imperative is thought to be a key feature of the third dimension of power (Culley and Hughey, 2008: 101-102).

Here one can argue the Foucauldian influence is recognisable through the idea of discourse. A discourse, in Foucauldian terms, is a way of thinking and speaking about some aspect of social life which shapes the possibilities for human action. Power is given special importance in Foucauldian approaches. Power is everywhere, not held by persons, but claimed in interaction through discourse with persons occupying complementary subject positions. Knowledge is also closely integrated with the concept of discourse in Foucault power analysis. The third dimension concerns the forces that determine what needs people recognise themselves as having; that is, how do I decide what my own needs are? (Lukes, 2005). In Foucault terms these forces can be characterised as discourses. This third dimension, then, concerns individuals' perception, cognition and self-view in such a way as to (a) shape the needs they perceive themselves as having; (b) determine the demands and requests the individual makes upon his/her self, others and society and (c) decide, at least in part, which issues may start to reach the political agenda (Kelly, 2006: 2119).

Furthermore, Lukes introduces and stresses the importance of the concept of latent conflict (Lukes, 1974). A latent conflict consists in a contradiction between the interests of (A) (those exercising power), and the real interests of (B), who are excluded. He asserts the conflict is latent because those subject to power do not express or even remain aware of their interests. This means that the interests of B are very difficult to trace, because those concerned either cannot express them or are unable to recognise them.

Critics of Lukes point to the difficulties of using the three dimensional view of power in empirical research (Hardy and Leiba-O'Sullivan, 1998; Lorenzi, 2006). First, the exercise of power may involve inaction rather than observable action. For example, due to financial constraint following a consultation exercise with the public, power holders may take a course of action other than what the public want. The point here is how to find a causal link between inaction and its consequences. Some of the questions that might arise here include justifying the claim that the powerless, the poor and excluded would have thought and acted differently, and specifying the ways or mechanisms in which those exercising power acted or abstained from acting in order to prevent the excluded from doing so. In order to gather evidence to support the claim that an apparent case of consensus is not genuine but imposed one must investigate inaction, consider structural and institutional power, and investigate ways in which demands are prevented from being raised.

The second problem of Lukes' three dimensional view is how to identify the process or mechanism of the alleged exercising of power. The exercising of power may be unconscious. This may be the case where A exercises its power over B yet remains unaware of its consequences. In this case, there is an exercise of power only where A could have discovered the consequences of its behaviour.

Third, power may be exercised by groups or institutions. This entails a crucial question: how and where does one draw the line between structural determination and the exercise of power? However, Lukes refuses the conceptual assimilation of power to structural determination. Within a system characterised by total structural determinism which is essentially a one dimensional view of power, there would be no place for power. Power, Lukes claims, is about alternatives, and that to identify a given process as an exercise of power is to assume that within the process lies the possibility to act differently. This holds for individuals as well as groups or institutions. His conclusion is that locating power is to fix responsibility for consequences that flow from the action, or inaction, of certain specifiable agents (Kernohan, 1989).

The different notions and functions of power are reflected in the arena of public and community participation. If we use both Foucault's and Lukes's concepts of power, then it is possible to explain public participation in terms of power relationships. According to Foucault, power can be used positively; accordingly the exercise of power within the context of public participation may also result in a positive outcome. Furthermore, as power does not emanate from one single source according to Foucault, individuals and certain groups do not always dominate the direction of the decision making process. As checks and balances could be put in place and the source of power varies, it is difficult for power to become concentrated into the hands of few, hence power could be used as a productive rather than repressive tool to accommodate the needs and views of different groups. Therefore, the question is more to identify who benefits from the decision making process through the participation of the public or the communities. The most important issue to be considered in public and community participation is the evaluation of both the process (where the actual exercise of power matters) and the outcomes that results from the process.

On the other hand, when we use Lukes' approaches to power to explain public and community participation, we identify at least three important issues. Firstly, many of the public and community participation exercises are subject to the quality and quantity of resource availability which determines the direction of their results. In most of the time material, technical, financial and human resources are held by power holders; this means that the direction of the public and community participation exercises tends to be influenced by the resource holders. Secondly, the powerful could be selective in deciding who is to participate or not to participate and issues to include and exclude (Rowe and Frewer, 2005). This means that the powerful will be in a position to control the rules of the game. Hence the result of the participation agenda obviously favours the powerful. The third aspect is more complex as the influence of the powerful is about shaping the perception, cognition and preferences of situations and events. Here the participation is carried out in the environment where the real interests of the public and communities are influenced by some other ideological, political or cognitive thinking. These are more embedded in emotional and current issues that instigate feelings on certain issues and conditions. These influences may not have a long term life span, but they have immediate impacts. For example, in a local political participation arena, current and populist issues could win the support of people who are directly associated with the issue.

Conceptualising empowerment

Empowerment concerns people's ability to claim and exercise power and frequently entails securing greater access to resources and the acquisition of skills to enable individuals and communities to assert control over their circumstances. Moreover, empowerment leads to building networks that promote participation and social action at individual and community levels. This notion addresses the issues of power and empowerment in relation to participation.

It was Paulo Freire (1973) who promoted empowerment as a social theory when he discussed how oppressed people could be educated to liberate themselves through local and community based initiatives (Parpart, Rai, & Staudt, 2003). Empowerment is not only about liberating oneself. According to Friedmann (1992) there are two important steps that lead to empowerment. The first is mobilising the poor to be a social force. This ultimately transforms into the second step of political power which is a step that enables participation in the decision making processes.

Authors like Banyard & LaPlant, (2002) describe the path that leads to empowerment as having three steps. The first promotes an interpersonal sense (of empowerment), or encourages participation. The second step builds community connections (networks) or integrates diversity. The final step promotes social action for community building or fostering involvement in the community. At the first stage empowerment practices within a community have led to changes from a situation of community diversity to one where there is unity of purpose in order to secure a common goal (i.e. integration). The second stage is that the intermediate step before community integration or community building is participation in or involvement with community activities.

The other approach to understanding the concept of empowerment is what Starkey (2003) categorises as the two distinct models of empowerment; these are: 1) The consumerist approach to empowerment – focusing upon individual change and control; and 2) The liberational model of empowerment - concerned with tackling unequal social structures. However, Traynor (2003:135) criticised the first approach by pointing out the contradiction of some notions of 'empowerment' which place emphasis on personal responsibility rather than considering the 'structural constraints on the life and consciousness of the individual'. Traynor argued that 'empowerment' can perpetuate an ideology that 'tells the individual subject that he or she is free while at the same time constructing the possibilities for thought and action'. Therefore, theory of empowerment needs to address the wider structural constraints.

However, I tend to agree with and use a different perspective: Hur (2006: 524) views empowerment as 'multidimensional' in that it has different meanings; it also functions 'at various levels, such as individual, group, and community'; 'is a social process because it occurs in relation to others'; and is 'an outcome that can be enhanced and evaluated'. Individual empowerment emerges when people attempt to develop capabilities to overcome their psychological and intellectual obstacles and attain self-determination, self-sufficiency, and decision-making abilities (Becker, Kovach, & Gronseth, 2004). Collective empowerment develops when people join in action to overcome obstacles and attain social change (Staples, 1990). Groups become empowered through collective action, but that action is enabled or constrained by the power structures that they encounter (Parpart et al., 2003).

The notion of "collective belonging" i.e. belonging to the social networks of peers, and an emphasis on autonomy while being part of the collective and social solidarity vis-à-vis establishment is one of the most frequently reported elements of collective empowerment in the literature. Boehm & Staples (2004:274) identify three components of empowerment: (a) collective belonging, (b) involvement in, and (c) control over organisations in the community. Involvement in the community means taking part in community activities or events that may lead to achieving change in or affecting the power structure in communities. For example, associations and neighbourhood community policing have structures to make decisions at neighbourhood levels. Such decision making at neighbourhood level is an example of empowerment through community engagement (Baillie et al., 2004; Zaldin, 2004) and coalition building (Boydell & Volpe, 2004).

Control over organisations in the community was considered as one of the critical components of collective empowerment (Boehm & Staples, 2004). Control over organisations means gaining forces to influence representative groups, plus ensuring the efficacy of those organisations. Control of organisations in a community can refer to group support and advocacy (Bellamy & Mowbray, 1999) and political control (Itzhaky & York, 2000; Zimmerman &Zahniser, 1991). Community associations are good examples of controlling organisations, raising their own funds to run them and taking decisions necessary for their causes.

Finally, the notion of community building was one of the critical components of collective empowerment. Community building refers to creating a sense of community among residents that will increase its ability to work together, problem solve, and make group decisions for social change (Fetterson, 2002; Mattessich & Monsey, 1997). Some authors describe it as social cohesion (Peterson et al., 2005) and a sense of personal freedom (Gutierrez, 1990). According to Gutierrez (1990), the goal of collective empowerment practices is to help communities develop the ability to change negative situations and prevent the recurrence of the problems that created those situations. This goal cannot be accomplished without the establishment of community building.

Hur (2006: 535) identifies the goals of empowerment for individual is "...to achieve a state of liberation strong enough to impact one's power in life, community, and society" and the goal of collective empowerment "...to establish community building, so that members of a given community can feel a sense of freedom, belonging, and power that can lead to constructive social change".

In the context of the conventional definition, empowerment must be about bringing people who are outside the decision-making process into it. This puts a strong emphasis on access to political structures and formal decision-making and, in the economic sphere, on access to markets and incomes that enable people to participate in economic decision-making. It is about individuals being able to maximise the opportunities available to them without or despite constraints of structure and state (Rowlands, 1995: 102). Empowerment also includes access to decision-making processes by which people become aware of their own interests and how these relate to those of others, in order both to participate from a position of greater strength in decision-making and to influence such decisions.

In summary, empowerment refers to the development of understanding and influence over personal, social, economic and political forces impacting on life situations. Individual concepts of empowerment refer to an individual's ability to make personal life decisions, and are related to constructs such as self-efficacy and personal competence (Schulz et al, 1995). Schulz et al (1995:310) further argued "Empowering organisations provide opportunities for individual growth and access to decision-making processes". Empowered organisations are those with influence over their environments, and the ability to affect the distribution of social and economic resources. The concept of community empowerment used in this framework considers communities to be made up of individuals and organisations. Individuals experience personal empowerment, through exercises to create changes within the community or through the influence they generate on public policies.

CONCLUSION

I have discussed some concepts of empowerment in a wider sense. What are some of the specific implications of empowerment in public and community participation? Generally, empowerment in the context of participation entails awareness building, consciousness raising and formal training activities that can lead to more assertiveness, determination and active involvement in the challenges and decision making process of local and national issues.

Most of the sub-Saharan African cities are transforming to the better. However, the political and institutional arrangements are in favour of those with access to legal, political and social resources which are used to enforce power and control over the powerless.

In the developed world, public participation may be defined at a general level as the practice of consulting and involving members of the public in the agenda-setting, decision-making, and policy-forming activities of organisations or institutions responsible for policy development (Rowe and Frewer, 2004:512). This suggests that people can be empowered through their meaningful participation in the decision making process. As said above, in sub-Saharan African countries and cities participation is much a tokenistic exercises. Rent seeking and corruption together with the lack of check and balance institutions resulted in a long term detrimental effect that produces suspicion and mistrust in the relationship between the power holders and the public. To avoid such suspicion and mistrust it is important to design a participation strategy that will make the public confident about the process of engagement in local urban development initiatives from its outset. The strategy needs to centre on empowering people and enhancing their capacities to participate through more flexible and appropriate methods including traditional network arrangements like Edir. The flexibility of the strategy is necessary to accommodate those people who may not be within the system of the mainstream society due to a variety of reasons. This is to say that, where people are not directly able to participate, there needs to be a system that will enable them to air their views, at least through their representatives or advocates. On the other hand, for those who are able to manage issues by themselves, direct participation could be appropriate with different means of support and empowerment mechanisms. As a result of such participation methods people might feel empowered to take greater responsibility for the care and maintenance of urban development initiatives in their neighbourhoods from the outset.

Nevertheless, they might not take responsibility and ownership if the urban development projects are imposed upon them with only minimal or non-involvement. One possible way to make public participation more effective could be through the empowerment of individuals who directly or indirectly have interest in the local initiatives particularly in newly emerging and expanding neighbourhoods and cities like Addis Ababa. The local partnership arrangement using different actors (including individuals) could be starting point for empowering local people in the decision making process. According to Atkinson, (1999:62) "Partnership and empowerment are central to the current phase of urban regeneration; however, the manner in which these terms have been defined and put into practice is frequently (and perhaps deliberately) somewhat opaque."

When partnership is discussed, one cannot rule out the inputs and expertise of agencies in support of the participation process and decision making by local (powerless) people. The contribution of local agencies, local municipalities and other professional groups is vital in order to educate local people about issues that are to be addressed by the development initiatives. They could also help in developing positive attitudes about the initiatives and in identifying the role that individuals and groups could effectively play and, eventually, in assessing the level of further support required by the community to make their participation more effective. In addition to the inputs of experts, independent voluntary and community organisations as well as civic associations can also have a real influence in empowering local people through practical support, particularly when there is suspicion and mistrust of the motives of the power holders. The pace of the decision-making process within urban development initiatives also needs to be sensitive to accommodate the different levels of understanding of the issues by individuals and community participants. McArthur, (1993:313) argued that: 'For external agencies this may often mean that slower progress has to be accepted in order to increase the potential for an effective contribution by local people and their representatives. However, this is an issue which cannot be separated from the question of resources'. Providing local people and communities with the opportunity to shape the strategy at an early stage helps to build trust and good relations between partners. It also provides encouragement for individuals to take participation seriously.

REFERENCES

Bachrach, P. and Botwinick, A., (1992). Power and Empowerment: A Radical Theory of Participatory Democracy, Philadelphia, Temple University Press.

Becker, J., Kovach, A.C., &Gronseth, D.L., (2004). Individual empowerment: How community health workers operationalize self-determination, self-sufficiency, and decision-making abilities of low-income mothers. *Journal of Community Psychology,* 32(3), 327–342.

Boehm, A., & Staples, L.H., (2004). Empowerment: The point of view of consumer. *Families in Society*, 85(2), 270–280.

Culley, M.R., and Hughey, J., (2008). Power and public participation in a hazardous waste dispute: A community case study, *American Journal of Community Psychology,* 41(1-2), 99-114

Dahl, R. A. and Tufte, E. R., (1974).Size and Democracy, Stanford, Stanford University Press.

Dahl, R, A,.(1989). Democracy and Its Critics, New Haven, Yale University.

Haugaard, M. (ed.), (2002). Power: A Reader, Manchester, Manchester University Press.

Haugaard, M., Lentner, H. H., (2006).Hegemony and Power, Lanham, Lexington Books

Holub, R. C., (1985). Remembering Foucault, *The German Quarterly*, 58(2), 238-256

Hur, M. H., (2006). Empowerment in terms of theoretical perspectives: Exploring a typology of the process and components across disciplines, *Journal of community psychology*, 34 (5) 523-540.

Kelly, B. D., (2006). The power gap: Freedom, power and mental illness, *Social science & medicine, 63*(8), 2118 -2128

Kernohan, A., (1989). Social Power and Human Agency, *The Journal of Philosophy*, 86(12) 712-726

Lukes, S., (1974). Power: A radical view, London, Macmillan.

Mills, C. W., (1956). The Power Elite, New York, Oxford University Press.

Newman, S., (2004). The Place of Power in Political Discourse, *International Political Science Review,* 25(2), 139-157.

Parpart, J.L., Rai, S.M., &Staudt, K., (2003).Rethinking empowerment: Gender and development in a global/local world. New York: Routledge.

Ron, A., (2008). Power: A Pragmatist, Deliberative (and Radical) View, Oxford, Blackwell Publishing Ltd.

Rowe, G. and Frewer, L. J., (2004). Evaluating Public-Participation Exercises: A Research Agenda, *Science, Technology & Human Values,* 29(4), 512-557.

Rowlands, J. (1995). Empowerment Examined, *Development in Practice,* 5(2), 101-107

Schulz, A. J., Israel, B. A., Zimmerman, M. A. and Checkoway, B. N., (1995). Empowerment as a multi-level construct: perceived control at the individual, organizational and community levels, *Health Education Research,* 10(3) 309-327.

Starkey, F., (2003). The 'Empowerment Debate': Consumerist, Professional and Liberational Perspectives in Health and Social Care, Social Policy & Society, 2(4), 273-284

Traynor, M., (2003) A brief history of empowerment: response to discussion with Julianne Cheek, *Primary Health Care Research and Development,* 4, 129–136

Young R. A., (1978).Steven Lukes's Radical View of Power, *Canadian Journal of Political Science,* 11(3), 639-649

CONCLUDING REMARKS

The book has brought together a variety of theoretical and practical insights from a number of perspectives that enrich our understanding of innovation, sustainability and ICT policies mainly focused on developing countries. Innovation and appropriate information and communication technologies as tools for sustainable socio-economic development are promoted in many developed and developing economies.

It is essential to put in place progressive policy regulations to make innovation and ICT effective and efficient. The involvement of dedicated government institutions is crucial, but the private sector should also be supported to bring innovations and ICTs to make them effective in solving the numerous problems and challenges all developing countries continue to face.

The partnership working arrangement between the public and the private sector is therefore paramount in order to bring innovations that can be supported by the introduction of appropriate ICT related systems in a sustainable manner. Admittedly, self-reliance is the ultimate goal for each nation, but more importantly, collaborating for a common good with regional and global bodies will ultimately produce a win-win situation. There is no doubt that any meaningful financial and technical support could be very relevant.

The national policy arrangement should therefore facilitate the amalgamation and adaptability of indigenous knowledge to new technologies in order to maximise the intended outcomes and minimise conflicts that might hinder the ultimate implementation process. This also brings the role of informal entrepreneurs to the front scene who constantly confront challenges with locally generated innovative solutions.

Innovations and ICT applications would not help the end beneficiaries, unless the governance of a country ensures transparency, accountability and the rule of law. Inclusive policies of Science, Technology & Innovation for

Sustainable Future, good governance is essential to maximise citizens' participation in the decision making process not only as an important means to sustainable development but also as a desirable goal.

Further highlights in the book include innovations and ICT development needs a well-grounded knowledge base. This brings the issue around knowledge management (KM) as very relevant and important to embed systematically in all development sectors. Some of the chapters in the book also indicated the relevance of KM as a systematic process that enables innovations and ICTs to yield sustainable socio-economic development.

The acquisition and development of technologies on the one hand and the applications of ICT in various sectors (e.g. higher learning institutions, banking, financial and health sector) as well as enhancing innovations on the other hand are also discussed in the book.

The book has covered a range of ideas that stimulate the reader to reflect on, as well as, to comprehend the challenges developing countries face in innovation and adaptation of ICTs. Obviously as challenges go side by side with opportunities, developing countries need to grab opportunities and translate them into more innovative technological solutions. Thus we believe the book has contributed to the ongoing debates on the wider innovation issues as well as to the relevance of ICT for the wider socio-economic development and concluded by recommending to have appropriate policies with particular reference to developing countries.

For many reasons, we also believe that the work on Africa is especially important. No continent has a more complex pattern of national boundaries or of ethnic and religious subsystems, interacting with sectoral systems than Africa.

The heritage of colonial exploitation and colonial divisions and the struggles for independence have left behind an extraordinary legacy which is further overlaid with the operations of multi-national and other international organisations. Ways have to be discovered to trace and understand these challenging multiple interacting cultural and institutional sub-systems to bring structural transformation by embedding knowledge, learning, innovation and competence building systems drawn from science, technology, engineering and mathematics. This volume two of the book series of GKEN can encourage debate as well as research on how Africa can genuinely reap the potential benefits, which science and technology as well as innovations can bring as this is the only way to achieve a meaningful development in a continent. It is long overdue that Africa is still lumbered with Afro-pessimism. It is time now not later to do away with the prevailing negative narrative and perception of Africa. It is important there is unity of leadership, knowledge, governance, institutions and values to remove all the lack of clarity and the lack of creative imagination.

As we move from MDGs to SDGs and science and technology and innovation is recognised as one of the 17 goals to realise a better future. So, this book on the STI for Global South is very timely and relevant and encourages all to engage with it.

Mammo Muchie, Mentesnot Mengesha and Amare Desta

NOTE ON CONTRIBUTORS

Edwards Adeseye Alademerin, PhD attended the University of Nigeria, Nsukka for his Teacher's certificate, Bachelor's, Master's and Doctoral degrees in Vocational Agricultural Education. His doctoral thesis was in the area of Programme Impacts Evaluation on cassava research and development in Southern Nigeria. He was a visited scholar at Aberystwyth University in Aberystwyth, Wales-Ache held positions as Head of Department of Agricultural Education, Deputy Director of Centre for Entrepreneurship and Vocational Studies and recently as the Director, Centre for Human Rights and Gender Education. Dr Alademerin is an Associate Professor in the Department of Agricultural Science of Tai Solarin University of Education, Ijagun, Nigeria.

Florida Alemayehu, MSc is a Lecturer, Project Manager and Software Developer by profession. Graduated in Computer Science Bachelor of Science degree from Mekelle University, Ethiopia, in 2006 and had a Master of Science degree in Information Science from Addis Ababa University, Ethiopia, in 2010. She has worked on projects that focus on mobile health, e-learning and technology assisted Instructional design development, in addition to teaching at Mekelle University for five and half years. Currently fully engaged on software development and project management mainly focusing on PHP and Java based projects.

Hasan Arab Ameri received his MSc in 1992 focusing on Materials and Metallurgical Engineering. His major specialties are in die casting industry. While he is an experienced engineer in manufacturing turbine parts, has gained many successes in business management.

Yesuf M. Awel is currently a PhD fellow at Maastricht University and UNU-MERIT. He was a research fellow at United Nations Economic Commission for Africa (ECA) in 2015.He was also a Lecturer in Economics at Mekelle University (Ethiopia).His research interest covers development economics, behavioural economics and applied econometrics. Yesuf holds MPhil in Environmental and Development Economics from University of Oslo and BA in Economics from Mekelle University.

Mengistu Bogale Ayele was a PhD candidate at Addis Ababa University (AAU) in the IT PhD Program (IS Track) who is currently defending his thesis and finalizing it. He is a lecturer at AAU School of Commerce in the Department of Accounting and Finance for the last 18 years. He is currently teaching courses such as Accounting Information Systems, Management Information Systems, Management of Technology and Innovation, Project Management and IT Auditing. His area of research include IT and Business Strategy, IT Governance, IT Auditing and IT Security.

Ali Reza Babakhan is currently a PhD student of Science and Technology Policy at the Progress Engineering Department of Iran University of Science and Technology (IUST).He gained his MS degrees from the Progress Engineering Department of IUST in the field of technology management, in 2014, and his BS degree from the Industrial Engineering Department of IUST, in 2012.His research interests include innovation policy and technology management.

Bedru B. Balana, PhD is a Research-Economist at the International Water Management Institute (IWMI), West Africa office (Ghana, Accra). Prior to joining IWMI he worked as a Lecturer and Assistant Professor of Economics at the University of Mekelle, Ethiopia (1995-2007) and Research Scientist (Environmental Economist) at the James Hutton Institute, UK (2008- Aug. 2014).Bedru obtained his Bachelor of Arts Degree in Economics (1994) from Addis Ababa University (Ethiopia); Master Degree in Economics (2000) from Punjab University-Chandigarh (India); and a PhD (2007) in Bioscience Engineering (Environmental Economics) from the Catholic University of Leuven (Belgium).

Dale Blahna is a research social scientist with the U.S. Forest Service's Pacific Northwest Research Station in Seattle, WA. He is also team leader for the PNW Station's Urban-Wildland Interactions Team, science lead for the

Urban Waters Partnership for the Green-Duwamish Watershed, and co-leader for the Green Cities Research Alliance. Blahna's primary research topics include recreation use of public lands, public engagement and use of social science in natural resource planning, and role of urban forests and parks in ecosystem sustainability.

Amare Desta, PhD is a Senior Lecturer and a Researcher in Management of Information Systems & Knowledge Management at the London South Bank University since March 2001.He is also an external examiner at London School of Economics and recently joined the University of the West Indies to serve an external examiner for Graduate School of Business. Dr Amare is a co-founder and chair of GKEN (Global Knowledge Exchange Network) since it is founded in 2011 and a Director of the Ethiopian Doctoral & Masters Academy. He earned his PhD & MSc from the LSE in Information Systems and Master of Education (MEd) from University of Cambridge, and BSc (Hons) degree in Computer Science from ARU, Cambridge, UK.

Haftamu Ebuy, MSc is currently a lecturer at Aksum University. He obtained his first degree in Information Technology from Mekelle University on July 2007 E.C. Afterwards he served as instructor at Arbaminch University for two years. Consecutively, after serving as in instructor in Aksum University for one year, he pursued and successfully completed his Master degree at Jimma University on November 28, 2013 in Information and Knowledge Management.

Joanna Endter-Wada, PhD is a Professor of Natural Resource Policy and Social Science in the Department of Environment and Society, Quinney College of Natural Resources, Utah State University. Her research focuses on integrating policy and social science contributions into the science and management of natural resources, diversifying perspectives used to understand human-environment interactions, and working at the interface of natural resource science and policy to promote a more equitable and sustainable future.

Melisachew Adane Ferede (MSc, MPH), is an International Public Health Consultant on Expanded Programme of Immunization (EPI)/ Polio Eradication Initiative (PEI) in WHO-South Sudan. Prior to joining WHO he has worked for more than 10 years as public health expert in Ethiopia. He holds a Bachelor's degree in Public Health from Alemaya University in 2001, MSc in Health Informatics from Addis Ababa University in 2009 followed by a Master of Public Health degree from University of Gondar in 2010.

Asebe Jeware has been trained for Oracle administration and performance tuning. He has got MSc in Software engineering from HiLCoE School of Computer Science & Technology and his first degree in Computer Science from the same college, Diploma in Banking and Finance from AACC. Currently he is working in the position of IT Manager at Vision Fund Ethiopia (WMFI) in Ethiopia.

Worku Jimma has obtained his B.Sc. degree from Addis Ababa University in 1994 and two masters' degrees, namely MSc in Aquaculture and Masters of Knowledge & Information Management, both from Gent University, Belgium in 2002 and 2004 respectively. Moreover, he obtained Masters of Religious studies from Nations University, USA in 2011. Worku has worked as a researcher based at different Research Institutes in Ethiopia and at the Laboratory of Aquaculture and Artemia Reference Centre, Gent University, Belgium. Currently, Worku is a faculty member of Jimma University and a PhD Candidate, specializing in Health Information Management at Tehran University of Medical Sciences, Iran.

Ermias Kebede earned his PhD in Business Administration, Masters in Business (MBus) Global Business Analysis, and BSc. (Hons) Physics with Business and Management from The University of Manchester (UoM). From October 2011 to July 2014, he was on the Thales Graduate Development Programme working in a variety of roles and business segments. At present time, Dr Ermias works in the UK Civil Service. Dr Ermias continues to engage in academia through his networks at UoM, delivering an annual guest lecture on the MSc Managing Projects course and through his work with GKEN and EDMA. His research interests cover public-private partnerships, acquisition systems, and transaction cost economics.

Mesfin Kifle earned his PhD in Software Engineering from Addis Ababa University. He has a rich experience in software development as programmer, system analyst, project manager and system consultant. Since 2004, he is teaching courses related to programming, software engineering, distributed systems, research methods to both undergraduate and postgraduate students. He is also engaged in community services by following a new collaborative approach to bring benefit to graduating class students, faculty and the community at large. Currently he is also acting as chair of the Department and project manager of the ISIMS project at AAU.

Linda E. Kruger is a social scientist with the US Forest Service in Juneau, AK. Her research focuses on community sustainability and resilience within a context of social, cultural, economic, and biophysical change; community tourism planning; and collaboration and partnerships, including challenges, benefits and innovative approaches to working together. Her recent work includes work with tribal elders and youth to document harvest and use of forest products, coastal climate change and impacts of climate change on tribal communities. She co-edited "Place-Based Conservation: Perspectives from the social sciences" (Williams, D., Stewart, W. and Kruger, L.) published in January 2013 by Springer Publishing.

Lemma Lessa, MSc is a Lecturer at the School of Information Sciences, Addis Ababa University, Ethiopia. He holds a M.Sc. Degree in Information Sciences and has joint the internationally supported PhD program in Information Systems at Addis Ababa University, Ethiopia. His teaching interest is on the social and management aspects of the design, adoption and use of Information Systems. His research interest is on the behavioural, socio-cultural, and organizational aspects of Information Systems in general; and the adoption and evaluation of e-Government in particular.

Robert J. Lilieholm, PhD is the E.L. Giddings Professor of Forest Policy at the University of Maine, and an Adjunct Lecturer at the University of Nairobi, Kenya. Lilieholm's research examines ways in which wildlands can be sustainably managed to promote a wide range of social, economic, and ecological goals. Examples include the modelling of alternative future development scenarios, and the creation of market-based approaches to sustaining protected areas and local communities. Lilieholm has authored numerous publications through funding provided by the National Science Foundation, the Ford Foundation, The World Bank and others. His work has been featured on CBS 60 Minutes, NPR's Marketplace, and Forbes.

Faramrz Lotfali is a practitioner who has received his BSc degree in 1990 focusing on Mechanical Engineering. For the last 25 years, Faramaz has worked on Turbine Design and experienced technology localization techniques, specially reverse engineering. Currently he is production manager of Tajrobeh Nour Company in Iran.

Mentesnot Mengesha, MPhil is a practitioner and an academic who has worked for the national & local governments, NGO's, NHS Trusts and academics institutions. His previous works include senior management positions for Menschen fur Menschen Foundation and Somali Refugees Care and Maintenance Programme in Western and Eastern Ethiopia respectively. Mentesnot has also served as a Non-Executive Director for acute National

Health Service Trust in London appointed by The Appointments Commission – UK. Mentesnot is one of the founders of the Global Knowledge Exchange Network.

Mehdi Mohammadi received his PhD in 2011 focusing on the issue of innovation systems. By now, he is assistant professor at faculty of management at University of Tehran. He teaches courses relates to technology and innovation management, Innovation policy-making, as well as innovation systems and learning in developing countries. He has conducted different researches in innovation and learning patterns, governmental policy-making in order to improve Iranian companies, he also published many papers in this regard. He is a member of Iranian association of management of technology (IRAMOT) and the director of Iranian award for management of innovation and technology (IRAMIT).

Mammo Muchie holds a DPhil in Science, Technology, and Innovation for Development (STI4D) from the University of Sussex. He is currently a rated DST/NRF Research Professor of Innovation Studies at Tshwane University of Technology. He is also TMDC Senior Research Fellow in Oxford University, UK, and Adjunct Professor in ASTU, Ethiopia. He is a member of the African Academy of Science and the South African Academy of Sciences. He is also the chair of the Network of Ethiopian Scholars that initiated also the GKEN (Global Knowledge Exchange Network).He is founding chief editor of the open access journal Ee-JRIF and the African Journal of Science, Technology, Innovation and Development (AJSTID).Since 1985, Prof. Muchie has produced over 365 publications, including books, chapters in books, and research papers in internationally accredited journals and entries in institutional publications.

Kingstone Mujeyi is a PhD student in the Department of Agricultural Economics and Extension at the University of Zimbabwe. His research interests are on informal sector dynamics and linkages with agricultural value chains. The focus of his research is particularly on the role of informal sector supply chains in meeting the technological requirements for appropriate agricultural mechanisation of smallholder agriculture in Zimbabwe.

Tamirat Fikre Nebiyu, MSc after his honorary BSc completion from Ethiopian Civil Services University, he joined the University of Stuttgart and obtained MSc in Infrastructure Planning. He worked as: transport modeller, planner, manager, and lecturer for several years.

Oluwole Ogunyemi, PhD is a Senior Lecturer in the Department of Agricultural Extension and Management, Lagos State Polytechnic, Nigeria and the Assistant Director (Records and Statistics), Academic Planning of the Polytechnic. He was a former Head of Department of Agricultural Technology in the Polytechnic. He holds First Class Honours B.Agric.Tech.Degree in Crop Production from the Federal University of Technology Minna, M.Sc. and PhD Degrees in Agricultural Economics from the University of Ibadan, both in Nigeria. He has attended a number of conferences and has some publications to his credit.

Mohammad Ali Shafia received his PhD in Industrial Technology of Tribology from Brunel University, UK, 1977.He is a Professor at IUST specialising in MOT, innovation and entrepreneurship. He is currently a member of Iranian MOT and Industrial Engineering Planning Committee for Iranian Ministry of Higher Education, a founder member of the National Iranian Productivity Organization, a member and Deputy Manager of Policy Making Commission of Industry for Government of Iran and a member of founding committees for NGOs like IE, Management of Technology, and Quality Management. He has published many national and international books and technical papers.

Mekbebe E. Tessema, PhD is a conservation social scientist and natural resource expert. He currently works for the New York based Wildlife Conservation Society (WCS) as Community Conservation and Livelihoods Coordinator in South Sudan. His research interest focuses on human and environment interaction, natural resource use and dependency, climate change impacts on rural resource dependent communities and protected areas, land use policy and governance in natural resource management, and rural livelihoods security and sustainability in biodiversity conservation. He also served as guest instructor and external examiner in different universities in the US and Ethiopia. Dr. Mekbebe has published his research works in various peer-reviewed technical reports and highly rated social science and ecology journals such as Society and Natural Resources and WIT Transaction on Ecology and Environment published by Routledge (Tayler & Francis Group) and WIT Press, respectively.

Richard Watson is a Regents Professor and the J. Rex Fuqua Distinguished Chair for Internet Strategy in the Terry College of Business at the University of Georgia. He is the current Research Director for the Advanced Practices Council of the Society of Information Management and a former President of the Association for Information Systems. In 2011, he received the Association for Information Systems' LEO award, which is given for exceptional lifetime achievement in Information Systems

Tsehaye Weldegiorgis is an Assistant Professor of Economics working in Mekelle University. He has obtained his BA in Economics from Mekelle University in 2004 and MSc in Urban Development with Distinction emphasis on Urban Social Development from IHS, the Netherlands in 2007. He has more than 12 years of experience in micro and macro quantitative and qualitative researches on livelihoods, finance, MSE, urban services and infrastructure. He has served Mekelle University as head of Quality Assurance and Director of Continuing and Distance Education; and he is currently working as head of Eco-Municipality Development Centre at the same institution.

Temesgen Abera Weseni is a doctoral candidate, lecturer, consultant, and researcher. He is a member professional associations, i.e. International (Association of Information Systems) and local (ESA and EPHA).His Ph.D. study focusing on PPP, e-Government, Knowledge Management, and e-Services. He has a number of studies published in IEEE, GKEN, and ESA. He has also served as a Web Chair for the IEEE Africon2015 and ESA2016 conferences. His research interests are Information Systems Development, Knowledge Management, Project Management, and ICT4D.He received the WSU's high achievement award from the University president.

INDEX

A

Absorption xxix, 8, 226, 241, 244, 245, 248, 249, 251, 252, 254

Actor-Network Theory (ANT) xv, 65, 67, 72

Africa Technology 13

Agrarian economy 3, 20

agribusiness 3, 4, 5

Agricultural mechanization xxii, xxiii, 3, 4, 5, 7, 8, 9, 10, 12

Analytic Hierarchy Process xi, xv, xxvii, 73, 163, 167, 168, 174

Aquaculture vi, 89, 92, 93, 96, 97, 334

Aquatic plants Scenario 94, 95

Auditing IT Governance 204

B

Biodiversity 28, 43, 95, 294, 337

Biotechnological innovations 93

C

Community-resource use linkages xxx, 295, 304, 306, 307

Contracting Process 109

Corruption perception index (CPI) xv, 51

D

Defence acquisition process 102, 119, 125

Defence procurement 99, 100, 104

Designing software architecture 207

Development iv, xiii, xix, xxii, xxiii, xxiv, xxv, xxviii, xxix, xxx, 5, 6, 7, 8, 10, 11, 12, 14, 18, 19, 20, 23, 24, 25, 27, 28, 29, 31, 34, 35, 38, 39, 43, 44, 46, 47, 48, 49, 50, 52, 53, 54, 55, 57, 60, 63, 64, 66, 67, 69, 71, 74, 75, 90, 92, 93, 95, 97, 104, 116, 117, 118, 131, 139, 145, 161, 162, 163, 166, 174, 175, 178, 180, 182, 190, 191, 195, 207, 208, 210, 213, 214, 222, 224, 225, 227, 228,

230, 233, 235, 236, 242, 243, 244, 245, 252, 253, 255, 257, 258, 259, 260, 261, 263, 264, 265, 266, 267, 268, 269, 270, 273, 274, 275, 276, 277, 278, 279, 281, 282, 283, 293, 313, 314, 321, 322, 323, 324, 327, 328, 331, 332, 334, 335

E

Economic benefits xxvii, 166, 167, 168, 171, 172, 174

Educational systems 7, 220, 226

Education type factor 281

E-government 63, 68, 73, 335, 338

Entrepreneurship xii, 55, 331

Experience and competency factor 280

F

Fishmeal 90, 93, 94, 95, 97

Foreign direct investment (FDI) vi, ix, xi, xvi, xxiv, xxv, 75, 76, 77, 78, 80, 81, 82, 83, 84, 85, 86, 87

Forest management xxx, 304, 307, 309, 310

Forest planning 307, 311

G

Good Governance xix, 55, 328

Growth and Transformation Plan (GTP) xvi, 62

H

Health Informatics xv, 150, 158, 333

Health Information Systems vii, xvi

Human Development Index (HDI) xvi, 51

Hybrid architecture xxviii, 208, 209, 210, 213

Hybridizing 208

I

IKM Framework xi

Indigenous people 4, 28

Indigenous technology xxiii, 6, 18, 19, 24

Informal sector xxiii, 4, 7, 9, 10, 11, 12, 13, 336

Information Systems (IS) vii, xvi, xxiv, 59, 73, 74, 142, 205, 206, 232, 233, 235, 236, 237, 238, 283, 332, 333, 335, 337, 338

Information Technology (IT) xvi, xxvii, 63, 74, 145, 158, 205, 206, 232, 233, 234, 236, 237, 238, 333

Innovation iv, xvi, xix, xx, xxi, xxii, xxiii, xxiv, xxvi, 4, 5, 6, 8, 10, 11, 12, 27, 30, 31, 35, 36, 38, 41, 44, 48, 49, 50, 51, 52, 53, 54, 55, 56, 66, 94, 95, 96, 133, 139, 181, 191, 233, 236, 241, 242, 243, 244, 245, 248, 251, 253, 254, 255, 327, 328, 329, 332, 336, 337

Innovation capacity index (ICI) xvi, 51

Innovations iv, xx, xxi, xxii, xxiii, 3, 4, 5, 6, 7, 8, 9, 10, 11, 12, 13, 27, 28, 31, 33, 36, 40, 41, 52, 53, 54, 55, 92, 93, 96, 241, 242, 243, 244, 245, 247, 250, 254, 275, 283, 327, 328

IT Governance vi, xi, xii, xxvii, 193, 194, 197, 199, 200, 201, 204, 205, 206, 332

K

KM Technologies 133, 140

Knowledge assets 29

Knowledge Exchange xix, 333, 336

knowledge gaps 50, 275

Knowledge Management (KM) vi, vii, xvi, xxvi, xxix, 131, 133, 134, 135, 136, 141, 142, 257, 258, 259, 261, 263, 265, 270, 271, 333, 338

L

Learning styles xxix, 247, 251, 254

least Developed Countries (LDCs) 226

Localization vii, xxix, 241, 242, 243, 244, 245, 246, 248, 249, 250, 251, 252, 253, 335

Location factor 280

M

Medical Informatics 158

Millennium Development Goals (MDG) xvi, 44, 62, 227

Multi-Attribute Decision Making (MADM) xvi, 163

Multi-criteria decision making (MCDM) xvi, xxvii, 163, 173

O

Open Source Software xvii, xxvii, 183, 191, 192

P

Participation vii, xxx, 313, 324

Pedagogical Practices vii, xxviii, 219, 237

Permit data xxx, 288, 290, 291, 295, 303, 306, 307

Power and Empowerment vii, xxx, 313, 323

Procurement Route 104

Public Participation vii, xxx, 313

Public Private Partnerships (PPPs) 60, 71, 72

R

Regulatory framework 12

Relational Contracting xxv, 103, 121, 122, 123, 124, 125

Research and development (R&D) 23, 50, 191, 242, 243, 245, 264, 331

Road infrastructure 162, 163

S

Scenario Planning 89, 91, 92, 97

Skills development 190, 230, 268

Smallholder agriculture 5, 6, 336

Social benefits xxvii, 166, 167, 168, 170, 172, 174, 192

Socialization (tacit to tacit) 140

Stakeholder Theory' (ST) 65

Sustainable livelihoods 28, 30, 311

T

Technological absorption capacity 252

Technological adaptation 252

Technological cooperation xxix, 245, 248, 251, 254

Technological innovations 4, 5, 6, 7, 8, 12, 13

Technology v, vi, vii, x, xii, xiii, xvi, xx, xxii, xxiv, xxvii, xxviii, 6, 13, 17, 25, 42, 45, 46, 57, 60, 63, 72, 74, 87, 131, 134, 135, 136, 137, 145, 154, 158, 174, 192, 205, 206, 214, 219, 222, 229, 231, 232, 233, 234, 235, 236, 237, 238, 245, 246, 247, 250, 251, 253, 254, 255, 278, 324, 328, 332, 333, 334, 336, 337

Technology and Innovation 13, 45, 255, 332

Technology transfer 75, 76, 85, 192, 241, 251, 255

The Ibrahim Index of African Governance (IBAG) x, xvi

The Indigenous Knowledge Management Framework vi, xvi

The operating system 178, 185

Toda-Yamamoto causality test 80, 82

U

Urban centres 4, 167

Urban settlement 313

W

Wealth Index (IWI) xvi, 51

Web-based learning 220